Writings on Israel and Entitativity

THE PEACEMATRIX™ Volume II

By Daniel Ben Abraham

ISBN: 978-1-961755-12-3

I0039337

"In individuals, insanity is rare. But in groups, parties, nations and epochs, it is the rule."

Friedruch Nietzche

"Entitativity" is the extent to which the group rather than the individual is the entity. In PeaceMatrix™ terms, it is, finally in humanity's development, the understanding of the ideological entities, the monsters, that cause mankind's wars. Do you see the monster's face in the clouds?

Contents

Foreword about the PeaceMatrix™

So, my quest for to implement my theoretical peace-building system to stop World War Three and bring about world peace has taken a bit of a turn southward. As they say, "it's not looking good for the good guys." And by good guys, I mean all of us who want long-term peace.

By now, I had hoped to be further along. Since Volume I was published, it has not garnered much attention. While some intellectuals find it fascinating, mostly people shrug their shoulders, and go back to more important things like scrolling social media, sports, movies, making money, and entertainment news gossip. Most people can't wrap their heads around it. I can't either. But global conflicts are escalating, and few are doing any real thinking.

In the PeaceMatrix™ Vol I, I talk about why humanity misunderstands our conflicts, why we get into them, and why we are tragically and gleefully headed to World War III and destroying ourselves unless we implement a new type of global peace-building system like the theoretical PeaceMatrix™.

Volume I is an introduction to the basic concepts of my invention for this new type of future, technological, metaphysical, peace-building tool for humanity. You should read it to understand how and why it will work, and for the necessary context of this, Volume II.

The PeaceMatrix™ book series is about my ongoing adventures and saga seeking to develop, test, and eventually implement this global system and bring it into reality. Increasingly, I'm thinking Elon Musk may be the only person with a mind open and expansive enough to be able to help me with this. Otherwise, I am here writing about my saga of trying to bring about world peace with a mere sketch of the Flux Capacitor sketched on a crumpled piece of paper instead of the final product, as we descend into the abyss.

As I was running experiments on the system, I discovered a strange anomaly in the lead-up to conflicts, that being, the parties don't behave rationally. Don't worry, there's more. I discovered that parties increase polarization, and with that, decrease rationality of judgment and problem-solving. As I sought to map out the rules of these phenomenon, why, and how, and patterns, I discovered what I call Entitativity Theory.

This book, Volume II, is the PeaceMatrix™ system's first real analysis of a few basic tools and concepts, including Entitativity, and it's a rough one. It is about potential, theoretical, analysis of largely of one principle, Entitativity theory, which is part of the development and analysis for H-Chain – "Cultures and Ideologies", which is just one of twenty-six (A-Z) starting questions or chains, of what may one day be a real, working PeaceMatrix™. The full list if Starting Questions is in Appendix A.

The system is not ready. I'm not ready. Humanity is not ready. But, we are playing near the water's edge of both civil and nuclear war, so all I can do is hope what I have is better than nothing. I'd give the quote that "you go to war with the army you have, not the army you want", but not sure how well it goes over for peacebuilding efforts.

Let's step back for a moment, and look at how the PeaceMatrix™ works, or at least, how is it theorized it will work.

The PeaceMatrix™ is a system of creating living, growing diagrams of any conflict between any number of parties, which serve as a global collaboration drawing board. The system's rules for diagramming the conflict guide the building of geometric models of the elements of disputes as puzzles to gain new understandings, and develop solutions.

When we solve the puzzle, we solve the underlying conflict. And, peace. The system develops broad to specific question chains down to every aspect of solutions necessary to overcome every barrier to peace, right down to the brain chemistry and metaphysical level.

The system, when fully developed and implemented, will revolutionize our human ability to understand, avoid, and resolve mankind's wars. The PeaceMatrix™ will one day solve the "failure of imagination" U.S. analysts blame, after the fact, for every single U.S. policy failure. The PeaceMatrix™ addresses conflicts before they start, before they boil over, before the CIA can assign someone to begin analysis on it, and before bureaucrats in Washington see themselves on the news and decide to divert attention to it. Certainly, before anyone with imagination miraculously works their way into these government bureaucracies, and miraculously works their way up into a position of power. Just kidding, that never happens.

At its top level, H-Chain asks the all-important question, "What about the parties' values, cultures, human natures, animal natures, beliefs, ideologies, customs, national, local, and tribal interests and other unknown forces and motivations is most important for understanding and peace?

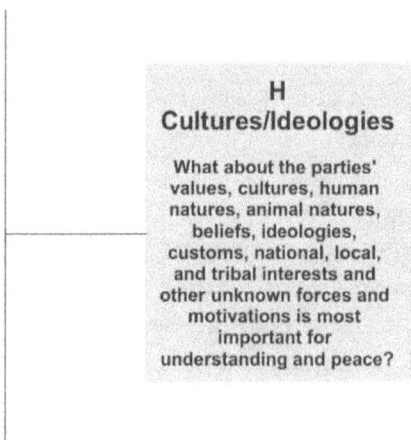

H
Cultures/Ideologies

What about the parties' values, cultures, human natures, animal natures, beliefs, ideologies, customs, national, local, and tribal interests and other unknown forces and motivations is most important for understanding and peace?

This very important chain seeks to understand each parties' ideologies, group mentalities, shared viewpoints, values, cultures, human natures, beliefs, views, emotions, customs, and national, local, political, and tribal interests. While our modern world treats these areas as separate, the PeaceMatrix™ combines these into one analysis chain as they all overlap. All these are another way of saying our motive, state of mind, common mind, and other compelling forces behind our actions - *why* we do what we do as a group, or individual members of the group. It is the nature and qualities of the spectrum of that party's ideology and collective mind.

Besides Entitativity, there are a couple of other major principles that I have discovered.

As it happens, one advancement in my progress towards world peace is that I now have a blog on the Times of Israel, where I have been publishing a blog about Israel, the current global and local conflicts, anti-Semitism and the Palestinian / Arab conflict.

I have also started my own blog, which I am developing, but tragically also far too slow at finding and building my audience. I might make a joke about the idiotic videos that get millions of views and followers, but I might start to cry.

These following articles introduce Entitativity, anti-Semitism, and the PeaceMatrix™ in that context. As they are interesting, I have included them in this book after the first few chapters on Entitativity.

These include the following:

-"The End of Moral Equivalence with Israel" - the factual basis for Israel's moral position after October 7th, including some of the facts necessary to properly understand and find paths to end the conflict.

-"The Metaphysics of Anti-Semitism" - a full chronology of 3300+ years of irrational anti-Semitic patterns in the human collective psyche, a new strategic definition, explanations of why a secure Israel is the key to global security, and why a "two-state solution" is currently metaphysically impossible.

-"The Trail of Mistakes to Nuclear War" - analysis of the ongoing strategic global mistakes by all major players, from Russia to China to the United States, which are today compounding towards nuclear conflict in everyone's disinterest.

-"Humanity is in its Infancy in Understanding War" – An introduction to Entitativity in this context, explaining how the threats we face supersede world leaders who are themselves victims of ideological forces, and how to understand and control those ideological entities to shape global events and avoid wars.

-"Adventures of a Young Mashiach Parts 1, 2, & 3" - a nearly-10,000-word piece looking at our problems from the prophesized Mashiach's perspective; why the U.N. has turned largely anti-Semitic and irrelevant, and an introduction to the PeaceMatrix™. As a new, replacement pro-Israel conservative world peace-building system, it can't turn anti-Semitic because Jerusalem would be its moral center, helping resolve all global conflicts and truly becoming "a light unto the nations". If Jerusalem could be involved as a mediator of disputes worldwide, the Jewish people would have a central role in global peace and security, and limit if not eliminate the emergence of harmful anti-Semitic ideologies.

-"Why Israel is NOT violating International law" - a short explanation addressing misguided allegations of "colonization", "genocide", "ethnic cleansing", "apartheid", "racism", & "occupation".

-"The Three Option State-Sinai-Relocation Plan" and "Entitativity Analysis: Palestinian Relocation" – These and other articles explain creating ideological momentum and the moral framework for global support for allowing Palestinians who wish to leave to relocate to the Sinai or other Arab states, as is the world's policy for all other refugee groups. Israel can barely explore these common sense ideas which also help Palestinians without angering allies and accusations of ethnic cleansing. With more opportunities and freedom for Palestinians who wish to leave, the other elements of the peace process may come together.

Disclaimer on several points.

First, I generalize viewpoints for the purpose of analysis regarding all parties and all our dear friends on all sides of these issues. The system would map out these distinctions and details from the general to the specific. In real life, it is very important not to take individual people as caricatures, although some are quite proud of their lack of self-reflection.

People are for the majority well-intended, trying to support what they see as justice. The problem is when we allow our communication systems to paint two separate moral frameworks instead of building one constructive and uniting set of values.

Most people on all sides are not evil, and of course, we have to define "evil". As we have different moral perspectives, different views of right and wrong, anyone acting from a different moral perspective is "evil" in a sense. Synchronizing those moral perspectives is the goal of a hopefully-well-developed Morality Chain, and its Solution-Chain counterparts.

Second, these writings may at times be critical of the media generally and make assumptions. There are many systemic faults within the nature of our competitive journalism field. This is not the focus of the book, and I am not stating that any individual journalist or specific media outlet is blameworthy. Journalists themselves are often unaware victims of the compulsions of their own ideologies. In short, they can't help themselves, though often they should.

Third, some may ask, if this is a peacebuilding book, why am I so hard on some opposing views? Shouldn't I be more neutral or balanced? The PeaceMatrix™ would map out all views on all sides, but I personally cannot write an infinite-paged perfect book with every issue mapped out as the PeaceMatrix™ would allow.

The book intends to examine ideology, not criticize individuals, and sees unhelpful individuals more as pawns in a game who have been influenced by their ideologies.

Not only does the PeaceMatrix™ encourage cooperation, but often in H-Chain analysis, the true enemy is not each other. We must better understand and improve how we deal with the effects our ideologies have on us, especially when our perspectives become skewed and group mindsets become malignant.

I also want to make expressly clear that I condemn all forms of hate. Obviously, as my life's work is about peacebuilding. And, the use of in-group versus out-group polarization for political gain that can turn malignant and lead to conflict, which is the underlying cause. When I discuss a potential conflict scenarios, it is a peace-building analysis of peaceful and constructive paths forward. My goal, and the landscape of global peace ultimately, should and will likely be one in which every beautiful and unique and peaceful culture is respected and cherished.

Perhaps the hardest thing to do to an irrational person is to prove to them that they are being irrational. Any of us who were ever married know this. As Mark Twain said, "It is easier to fool someone than to convince them that they've been fooled." Perhaps even more difficult is to persuade them to act against their raging emotions and compulsions, for the sake of a distant whisper of logic.

As the saying goes, "they know not what they do". But maybe we can tell them. And maybe they can see.

But, I will use the PeaceMatrix™ and try anyway. My hope? That Carl Jung was correct when he suggested, "Unless you make the unconscious conscious, it will forever dominate your life and you will call it fate."

What else did Carol Jung say?

"We need more psychology. We need more understanding of human nature, because the only real danger that exists is man himself. He is the great danger, & we are pitifully unaware of it. We know nothing of man, far too little. His psyche should be studied, because we are the origin of all coming evil"

That understanding, of how to manipulate the human psyche toward peace, individually and collectively, is the goal of my work.

Perhaps with PeaceMatrix™-style organization and mapping of concepts and facts, we can better understand and communicate what is happening. Of course, I am just using the diagramming for a very specific and limited purpose, not near the full potential scope of the system.

This book approaches escalating conflicts between ideologies in an organized and scientific way. And while the PeaceMatrix™ concepts are in their infancy, I think there's a part of all of us who want to believe that our ability to pursue peace may one day take giant leaps forward. So why not start now.

I don't believe war must be our fate. This book is written to try to prevent coming wars, by turning conscious our unconscious ideological escalation at the expense of our peace, so we can see what is happening clearly, so cooler and wiser heads can prevail.

The people on the other side are not the enemy. The enemies are the ideological entities that seize and override otherwise rational behavior with primal hive-mind emotions, enslave us, create group-think decisions, dangerously escalate, bias our decisions, make us make errors, break down the systems that have maintained peace, and eventually, give their human adherents no other choice but war. We must be smarter than *them*.

To help further these understandings, I have assembled most of my Post October 7th writings on the PeaceMatrix™ and Entitativity in the context of Israel's ongoing and expanding war, and Ukraine/Russia/NATO. For more, please subscribe to my Substack blog at www.danielbenabraham.substack.com and email list for updates at www.danielbenabraham.com and www.thepeacematrix.org , and follow me on Twitter at @thepeacematrix and @danielbabraham1.

Peace,

Daniel Ben Abraham

Daniel Ben Abraham's Quotations on World Peace

"The opposite of war is nuance"

Daniel Ben Abraham

"The anti-semitism that plagued Jews for 2000 years in every community will increasingly infect the whole world as the world becomes a community."

Daniel Ben Abraham

"If humanity could find a way to answer the moral questions it stumbles over, it would mean the end to all war."

Daniel Ben Abraham

"We are a light unto the nations, whether the world wants to see that light or not."

Daniel Ben Abraham

"Maybe the way to solve an unsolvable puzzle is with another puzzle."

Daniel Ben Abraham

"Maybe the way to make the world more peaceful, is to make it more Jewish"

Daniel Ben Abraham

"Anti-Semitic Europe embraced Nazism, then after World War II pretended it learned its lesson, and simply went on to embrace anti-Semitic Islamic extremism, proving it learned nothing."

Daniel Ben Abraham

"Anti-semitism appears incurable except in individuals who have risen to a level high enough to accept the mysteries of Jewishness"

Daniel Ben Abraham

"Israel must be a state because the world has lost moral perspective, and will destroy itself without it."

Daniel Ben Abraham

"The future of war is not missiles nor drones nor bombs, as much as it is strategic understandings of how to get enemies to moderate each other."

Daniel Ben Abraham

"Making sure wars are endless appears to be an art form" (Regarding Biden/Obama in 2024)

Daniel Ben Abraham

"If Satan were running things, trying to encourage and fuel both sides into escalating war, he couldn't be doing a much better job." (Regarding Biden/Obama in 2024)

Daniel Ben Abraham

"Humanity's wars are not over land nor resources nor religion. Wars are not caused by individuals, nor soldiers, nor even leaders for the most part. Wars are caused by the moral questions that humanity stumbles over. When we cannot answer these moral questions, we begin to polarize on opposite sides of them in group collective mindsets. Each side becomes increasingly ideological and irrational, breaking down the systems that maintained peace, believing it is defending itself from a worse threat in the other, until war is the only option left."

Daniel Ben Abraham

"The solution to both politicized anti-semitism and all conflict amongst man is the same – a new global peace-building system based on and in Jerusalem that continually resets the world's moral compass by solving all global conflicts, maintaining a Talmudic-style debate to keep evil ideologies from rising."

Daniel Ben Abraham

"The Jewish people are not "the chosen people" for any privilege other than the privilege of the suffering necessary to bring the world back to a moral path. So please don't make it any harder."

Daniel Ben Abraham

"What's more powerful, a bomb, or the ability to turn the world from anti-Israel to anti-Iran in one day?"

Daniel Ben Abraham

"Those on one side won't be outraged by crimes of others on their same side, because their purported moral compass's logic is subservient to emotional tribalism. The difference between other moral perspectives and the Jews', is that civilization is build on and inseparable from the Jews. As war is a battle for moral perspective, the secret to ending all war against mankind is a moral perspective grounded in Judaism as its compass."

Daniel Ben Abraham

"God, Because when you pray for help with your problems, who else would give you more problems to help you appreciate what you have."

Daniel Ben Abraham

"Moral perspective is relative. If you identify tribally with an in-group, that subconscious collective hive mind will overpower your ability to reason right from wrong objectively."

Daniel Ben Abraham

"A true judge's rulings are enforced simply by being correct."

Daniel Ben Abraham

"The Palestinian problem in a nutshell is that the "Palestinian cause" is actually two ideologies, one that wants to live in peace, and a second that wants to conquer Israel. The problem is they are intertwined, and when the world feeds one, they feed the other. And they run down the heart of each Palestinian. The metaphysical solution is the universe of ideas that helps the first while necessarily and certainly hindering the second."

Daniel Ben Abraham

"War is a battle for moral perspective, controlled not by reason but ideology, which is in turn controlled by primitive in-group versus out-group polarization, which in turn behaves according to the properties of an ideological entity that mimic a living organism. In the Stone Age, if a member of your tribe killed a member of another tribe, it was fine; and if they killed a member of yours, it was infuriating. A mugger would have as much right to kill his victim as the victim has a right to defend himself if your moral persecutive were rich versus poor instead of criminal versus law-abiding."

Daniel Ben Abraham

What is the PeaceMatrix™?

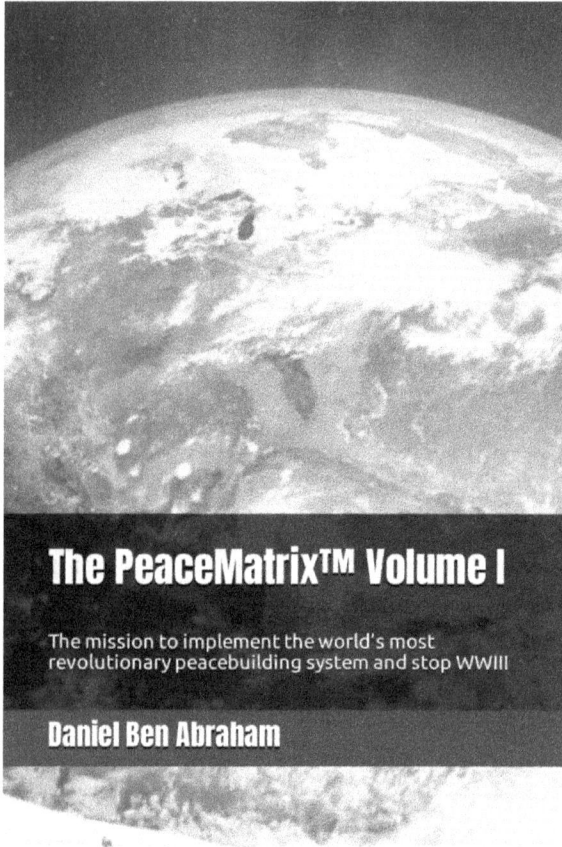

The PeaceMatrix™ is a new theoretical global peace-building system conceived by Author Daniel Ben Abraham designed to literally help create peace on earth, solve every potential and actual war between every nation, tribe, culture, religion, political group, and ideology, and prevent WWIII.

Just as in 1939, Albert Einstein wrote a letter notifying President Roosevelt of the theoretical possibility of the development of atomic bombs, I am writing now to inform of the theoretical possibility of a communication system that could bring about **global world peace**. It would expand human consciousness to prevent civilization-ending wars, and for mankind to finally have a superhuman perspective, and be able to think rationally.

We have seen movies with countless sci-fi inventions, but nobody has ever even conceptualized a system to end all wars. Now, maybe I have.

How will it work? As a neurodivergent myself with a background in international law, I've spent four years developing what I call "The PeaceMatrix™". The PeaceMatrix™ is a theoretical modeling system that guides creation of a geometric diagram of every viewpoint of every conflict. It serves as a drawing board for all of mankind to communicate, cooperate, and collaborate on in a solution-oriented manner. Since asking the right questions is what advances humanity, the PeaceMatrix™ is a map of outstanding questions of the conflict in a living, growing puzzle. It extracts and hyper-organizes all of humanity's best ideas combining understandings of human nature, philosophy, history, law, geometry, psychology, neurochemistry, and metaphysics. As the world interacts with the system, we each solve tiny micro-bits of the puzzle, and when we solve the puzzle, the underlying conflict is solved. We use one puzzle, to solve another puzzle. No more wars.

The system may be implemented through mediation and negotiations, as a global "chess game" debate, or as the organizing back-end content management system behind a social media platform's algorithm, possibly facilitated by Artificial Intelligence. Just as social media can bring out the worst in some people, properly tweaked, the right communication system can bring out the best. Instead of polarizing communication bubbles, it would turn all human dialogue into a constructive solution development system with common sense that awakens people from unhelpful ideologies like wokeness, emotionalism, and tribalism.

Our problem now is the world is arguing over the wrong issues - the inflammatory ones and not the constructive ones. Every point is missed like ships passing because we have no central organization system for the debate, no common ground, no way to advance the discussion constructively through the ocean of unhelpful information and emotions. We can't agree on definitions, moral perspectives, or other elements that make up our conflicts. Millions go to war without being able to even ask the other side a simple question like, "what do you want?" Being able to categorize the elements of disputes is the key to solving this, so we are finally comparing apples to apples.

The final frontier is conquering man's primitive human nature, so humanity can find the peace to accomplish all our other goals, and reach the stars. This may be one of the greatest, most mind-boggling discoveries, inventions, and adventures ever, and a new dawn for mankind.

Basic theories behind the PeaceMatrix™

The reason Israel is being raked over the coals in the United Nations, by the Security Council, the General Assembly, the Human Rights Commission, the International Court of Justice, and the International Criminal Court, is two-fold.

a) these systems have become corrupted by anti-Semitic ideological motivations that override human rationality like all anti-Semitism, and,

b) we haven't come up with a better system to replace it.

But a single divine idea can be more powerful than the anti-Semitic United Nations, and every anti-Semite in the world.

The worst part of having invented a global solution to end Israel's wars, America's wars, and all of humanity's wars, is, most people do not understand what I am saying.

For those, here is a summary of theories behind my ideas for a new world peace-building system, the PeaceMatrix™, to replace the anti-Semitic United Nations with a better system based on Jerusalem's core values, to bring about world peace.

1. Humanity gets itself into major wars every few generations, and our relatively peaceful lives are illusions.
2. Those wars will turn nuclear now that we have entered the nuclear weapons age, and permanently devastate mankind on our current course.
3. The propensity for war is embedded in our most primitive aspects of human nature, far more powerful than our logical and rational expert analysis which proves to be wrong on nearly every war.
4. World leaders and the systems they have built fail to correctly address the underlying ideological nature of war, which will be more powerful than leaders' attempts to avoid wars.

5. The United Nations and our current peace-building systems are faulty, anti-Semitic, focused on attacking and weakening Israel, have lost all moral perspective increasingly honoring the terrorist-led Palestinians with statehood, and is unable to prevent World War Three on our current course.

6. The fundamentals of our world; our nation-state system, our top-down leadership, our entire system of international law behind these likewise have material flaws, are turning anti-Semitic, and will fail to prevent World War Three, and permanent nuclear war amongst mankind.

7. The current international systems have become corrupted over time as all rules-based systems do, have become ideologically led-by anti-Semitic ideologies and those who do not understand them, and cannot be fixed.

8. Today's technological communication and media systems amplify the factors that lead humanity to war.

9. Peace is the most important goal of mankind, as without it, we cannot achieve any of our other goals

10. A new and completely different type of peace-building system to enhance or replace current systems is needed to address these problems.

11. The system must work on the individual level, instead of allowing ideologies to rise, and hoping world leaders voted in or empowered by those same ideologies will resolve them.

12. By categorizing the elements of disputes, we can exponentially progress in our understanding of DNA of human conflicts.

13. The broad categories of the PeaceMatrix™ contain every cause of every conflict among groups by encompassing understandings of; parties, term definitions, histories, current situations, parties' wants, why wants conflict, communication, ideologies, writings, moral perspectives, leaderships, obstacles, etc.

14. The solution categories of the PeaceMatrix™ contain every avenue for their resolution; ex: culture building, morality synchronization, communication building, common interests, all peace scenarios and solutions, overcoming their obstacles, etc.

15. Continually developing questions and their derivative questions from these in strategic chains will isolate the keys to resolving all disputes.

16. Humanity's best ideas for peace can be extracted from debate, continuously built one element of a solution upon another, manifesting their solutions through implementation.

17. Humanity can replace majority rule, ideological rule, and mob rule with a wiser system.

18. A Talmudic-style Socratic global debate that develops these question chains and solution chains presenting all sides will find and develop all solutions to all of humanity's wars and change the nature of man by making us more rational, taking us to a divine perspective.

19. That the solution to the "what is truth" problem is the diagramming of all perspectives because the rational debate will take the dispute out of the realm of the ideological.

20. Constructive questions addressed en masse by humanity will yield metaphysical dividends by opening worlds of infinite solutions.

21. Such a system will be infinitely better at predicting, preventing, and resolving disputes than the combination of our current inflammatory media and imperfect political processes.

22. Such a successful system unleashed will increasingly enhance and then replace the United Nations and current, faulty, and anti-Semitic systems.

23. Such a system based on Jerusalem's core values will prevent the rising of anti-Semitic ideologies which increasingly diverge from the world's moral source, turn evil by definition, try to conquer the world, and threaten human security.

24. Developing and implementing this system will allow Jews to maintain a debate that continually resets the world's moral compass with Judaism, preserving global peace and security for all peoples and cultures.

25. Such a system will help the Jewish people fulfill our purpose to be a light until the nations.

Any questions?

In case you're wondering what you missed in PeaceMatrix™ The PeaceMatrix™ Volume I and aren't going to read that first, here are the chapters:

- Summary
- A note from the author
- Part 1 – Introduction - Welcome to the party
- PeaceMatrix™ - The scientific approach to peace building.
- The PeaceMatrix™ explained backwards
- Why is the PeaceMatrix™ needed?
- Faults with all our "peacebuilding systems"
- Conflicts are brewing worldwide
- Our peace systems are increasingly strained
- Judeo-Christian values and Anti-Semitism
- Where did humanity go wrong?
- Crowdsourcing ideas for peace
- A social media platform for implementation
- Nerds unite
- What's so good about a diagram?
- Categorizing the elements of a dispute
- We are arguing over the wrong things
- Winning over whataboutism
- The diagram guides constructive communication
- Minimizing emotions, tribalism, cognitive dissonance
- Digging deeper into why we disagree
- An organization system for "outside-the-box" thinking
- Better, more diverse ideas
- Why the PeaceMatrix™ resolves "impasses"
- The question-based system
- Our conflicts are largely unanswered questions
- Ask better questions…get better answers
- The Apple.
- The Russia-Ukraine conflict as unanswered questions
- Metaphysics, finally!

- So, what is the PeaceMatrix™, and how does it work?
- The PeaceMatrix™'s moral framework
- Advantages over other systems
- Why the PeaceMatrix™ is revolutionary
- Shall we play a game?
- Part 2 - PeaceMatrix™ Starting Questions:
- Part 3 – First Active PeaceMatrixes™
- Part 4 – FAQ and Challenges Going Forward
- Conclusion
- APPENDIX A – The Starting Questions in more detail

Introduction to Entitativity

In December 2021, in the Lavool village in India, wild street monkeys banded together in a species war killing every dog in the town, after one dog killed an infant monkey. In war, humans aren't much different.

When humans form mobs and riots, they also lose their rational individuality, and are taken over by a "hive mind"; irrational, primal. This happens to world leaders also, not that they would tell us, even if they understood the phenomenon at play. But what if we…

Mankind is in its infancy in understanding the true reasons why we go to war, and it is about time we grow up. Humanity's wars don't happen for the excuses we are told. They are not caused by disputes over land or money or borders or resources. Individual soldiers do not even decide to go to war. But what if neither do our leaders?

Mankind cannot prevent wars because as we approach war, we become irrational, leading to errors in judgment that fuel the conflict. Wars happen because our analysis for understanding conflicts is flawed. Our fundamental underlying misunderstandings are believing we are rational and making our own decisions. If we got past those falsehoods, we would see that wars are ideologically driven.

The true puppet masters shaping humanity's conflicts, soon nuclear conflicts, and even a brewing second American civil war, are our own primal group ideological adherence. Wars occur because ideologies become powerful, and turn adherents, including our leaders, emotional, biased, and irrational. Whether in domestic politics or international wars, ideologies form group-think mentalities and make bureaucratic and ideological decisions that no individual is responsible for. Such hive-mind decisions on opposite e sides of conflicts escalate against one-another, with each side believing they are justified defending themselves from a worse threat in the other. The ideology's increasingly influenced adherents break down the systems that maintain peace, the treaties, the relationships, the norms and customs, until eventually the only option left is war.

One of the most irrational ideologically-spread mental illnesses that lead to conflict is anti-Semitism. This insanity that gripped civilizations and localities for over 3,300 years grew and spread until it compelled attacking, killing, expelling, burning, drowning of Jews until they were all exiled elsewhere. But now that we have a nation to defend ourselves, all out conflict is certain absent an anomaly as the world increasingly polarizes against the Jews.

But while irrational, there is a method to the madness. The system has rules that govern its growth and spread and genocidal compulsions and political influence. The good news, is that in properly understanding anti-Semitism lies the secret to preventing all wars among mankind. For if we could understand that conflict-prone disease of human irrationality, we might unlock them all. And we as Jews could be "a light unto the world", if the world doesn't murder us all for trying to be.

Wars happen because ideological entities possess the collective unconscious of group hive minds, override rational decisions, and compel adherents to break down the systems that have maintained peace until the only option left is war. This is the same in wars between nations, civil wars, religious wars, tribal wars, or political disputes, embedded in the human primal psyche since prehistoric times.

So if ideologies don't act according to their proclaimed values, what are the rules and nature of their behavior for hurling humanity to conflict every generation or two? Why are our best formulas and equations by which we attempt to calculate the paths to avoiding war always incorrect?

The answer is, there are individuals, there are leaders, and there are ideologies. The entire international system is set up to resolve disputes between leaders. And that is incorrect. Leaders are themselves controlled, unconsciously, by their groups' ideologies. In fact, the ideologies put those leaders into power. And, as we see in American politics, systems will violate their own rules to replace leaders who confront its ideological and systemic norms.

Our ideologies are not only separate players with their own seat at the negotiating table we've ignored. They are alive. Ideologies "possess" collective hive minds in behavior that instead mimics the power and growth dynamics of separate living parasite organisms. Ideologies have separate interests from the adherents, override the interests of the adherents, and override the proclaimed values of the ideology. Ideologies in our collective unconscious group psyches act as separate living entities and are the puppet-masters of humanity's wars. They gain power through the emotions of in-group out-group polarization, and override mankind's ability to rationally build bridges to peace. This understanding, and its derivatives in understanding how to manage these power dynamics are the keys to ending all wars amongst mankind, and world peace.

"Entitativity" is the measure of the degree to which an individual versus the collective group is the sovereign entity. It is therefore, an element affecting the distance between peace and war in any conflict. The PeaceMatrix™ Entitativity theories here have the solutions, the missing pieces, to make all of mankind's conflict analysis formulas work correctly. And if properly implemented, they are the ability to predict, avoid, and correct all war; the beginning of the end of all war amongs mankind.

The PeaceMatrix™ is a conceptual global peacebuilding system intended to utilize these understandings and resolve every conflict on earth. It is a mere concept in its infancy and not ready for such a challenge, but nor are we ready for the consequences if we fail.

Since the Ukraine war, we see the globe polarizing across the ideological divide of the conflict. And, we see step by step the downward spiral of breaking down the systems that maintained peace. Russia is suspending participation in the New Start Treaty, the last existing nuclear cooperation treaty with the United States[1]. Russia also announced stopping cooperation on the International Space Station, and instead began more military cooperation with Iran and North Korea.

We see the Biden Administration making a group-think decision with NATO toward a war in Ukraine that no individual is actually responsible for. It's "NATO" making decisions, as if NATO were a person. Meanwhile, Russia prepares the world for long-term economic war to weaken global U.S. influence, which is our greatest peace maintenance system of all.

We also see the simultaneous breakdown of peace maintenance systems with China.[23] We see China rejecting the Permanent Court of Arbitration at the Hague's ruling regarding the South China Sea. We see escalations, territorial expansion, security expansion, and counter moves by China's Asian counterparts from Taiwan, to Japan, to South Korea, to and while each separate move seems like a smart strategic move by itself, at one point, the only escalation left will be war.

We see the same in the Middle East conflict. Palestinians support Hamas who burn babies alive for being Jewish, and commit the most inhuman acts imaginable. Arab states that control 99.6% of the land in the Middle East and enormous wealth refuse to take in any Palestinian refugees because forcing them to stay in Gaza and suffer for the Palestinian cause for the genocidal goal of destroying Israel seems more moral to them. The world increasingly demands that Palestinians who wish to destroy the Jewish state receive a state with the means to do it. The world polarizes around this demanded impossibility, until war is the only option left.

---

[1] https://abcnews.go.com/International/live-updates/russia-ukraine/putin-suspends-usrussia-nuclear-treaty-97350698?id=88905005
[2] https://www.scmp.com/news/china/diplomacy/article/3231933/taiwan-withdraws-central-american-parliament-after-vote-favour-beijing
[3] https://theconversation.com/the-us-and-china-may-be-ending-an-agreement-on-science-and-technology-cooperation-a-policy-expert-explains-what-this-means-for-research-212084

In each case, each side's apparent best judgment backs the other into tighter and tighter corners blind to the true causes of humanity's conflicts.

But imagine we could map out the ideological elements on both sides that break down our most necessary peace-maintenance systems. And imagine we could show, in a non-hateful, but new, creative, and constructive way, the factors increasing and decreasing the effectiveness of our peace-maintenance systems. Just as it takes three points to prove a straight line, by plotting out the key patterns, across individuals and circumstances, we may better both see what is happening, where we are headed, and how to alter course. It would be seeing an entirely new map of the conflict, on an entirely new level. Like finally having night vision in the dark.

With such a tool, we may finally rise above our savagery to a higher perspective. To conceivably, for the first time in mankind's history, use this new tool, and go from outsmarting each other straight into war, to outsmarting war.

Getting back to the Lavool monkeys in India

In response to the killing of one of their young, monkeys banded together in mobs, dragged every dog in the town up trees or to the tops of tall buildings, and then threw them off, killing them, or should I say, murdering them, in an apparent "species war." Murder, as opposed to an accident or mere homicide, is the killing with "malice aforethought."

By the way, the law presumes that children under seven years of age are unable to form the intent to commit crimes because every crime requires not only a wrongful act, but an accompanying wrongful mental state. More on that later.

As an example regarding mental state in criminal acts, for human adults, legally, the same act of running over a pedestrian with your car could be an innocent accident if they darted out in front of you giving you no time to react, negligence if you weren't paying attention and didn't stop in time, recklessness if you were drinking, or murder if you actually intended to hit the pedestrian. Intent matters.

Regarding the Lavool monkeys, I'm pretty sure I know what question you now expect me to ask. You think I'm going to ask how all the monkeys coordinated. How they all worked together and how their shared mindset functioned socially and biologically to commit this mass murder.

Or you might even expect me to ask how different these monkeys are than us humans. That's not what I'm going to ask you. Though, chimpanzees and bonobos have about 99.6% genetically equivalent DNA to humans, just fyi.

What I actually want to ask you is far more frightening:

Did the monkeys have a choice?

Did they?

Or, were they helpless to follow the group mob mentality which brought them to organize, cooperate, and murder. And when I say mentality, I mean a group mob mentality – one so powerful the individual rational mind was an unknowing and helpless slave to it. A slave, whereby, the mass murder was, by unstoppable compulsion, the necessary and only action to be taken.

If the monkeys didn't have a choice, who did? Were they standing up for values? Or, did a quasi-living ideological entity use the killing fury and mob's emotional energy to gain power and come to life to control those monkeys? How did hundreds of monkeys click into violent group behavior, and who flipped the switch?

And could humans be similarly situated when we go to war?

Or, can humans break free from shared mindsets that exist on such primal levels? Look at our history-long list of wars and genocides before you answer. The people who committed those atrocities were allegedly "rational" people just like us, compelled to do so, as you would have been in their shoes. We must get off our high horse and face reality.

It gets scarier.

If the Monkeys were merely rationally upset, explain the following:

After the Lavool monkeys killed literally every single last dog in the town, the monkeys began targeting small human children. Now the plot really thickens.

The ideology not only took control of the monkeys to commit mass revenge against dogs, but after, it took on a life of its own and wanted to keep living (and killing) after the ordained task was done. Rationality be damned.

Hitler didn't become *the* Hitler because he had bad ideas. He became so because he had the uncanny ability to speak, so hypnotically, so compellingly, that such primal emotions overtook the crowd. We as humans are susceptible to "catch" and share the emotions of our peers in a phenomenon known as "emotional contagion." That's why when we watch movies, we feel what the character on the screen feels.

Because he spoke to people in crowds, and the crowds could be heard on the radio and seen on the television, the group hive mentality was triggered, and took control of the German populace. Hitler made the group mindset contagious, and enough Germans caught it that they were compelled to carry out his plans, each justifying it in their own rational minds and in their own personal unique way.

And we are no different today.

Our predispositions can be ideologically triggered, becoming emotional basis for a hive mindset, and motivation of entire groups.

On October 7, 2023, mobs of the Palestinian group Hamas, funded by Iran, broke into Israel, and not only burned babies alive and decapitated babies. They not only killed children in front of their parents and vice-versa. They not only tied parents to children and doused them with gasoline and burned them alive together. They even had an obsession with targeting the genitals of their male and female victims with knives and guns and blunt instruments in a way reminiscent of how the most primally-enraged chimpanzees attack their enemies.

The most striking part for those who watched real footage of October 7th filmed by the terrorists themselves, is their utter jubilation at committing these acts. They were ecstatic. And when one terrorist called his parents and told them by phone he had just killed ten Jews with his bare hands, his parents enthusiastically congratulated him. Obviously, part of the same hive mentality.

And before Israel even responded, anti-Semitism spiked by many-fold, and crowds of millions of supporters on the pro-Palestinian side gathered in cities around the world, mostly in Europe and the United States interestingly.

But were they there to show support, or there to feel the power of their connected primal hive mind in the crowd?

We see irrational hive-mind mob conduct in the context of American political dispute as well, though of course, so far on an entirely different level. Mobs of people tore down statues of founding fathers all over the country during the George Floyd riots. The mob, of course driven by hive mind desire for power and chaos they mistook for justice, had no interest in the fact that the founding fathers are responsible for our, and their, freedom today.

A lawyer for PBS resigned after saying Trump-voter's children should be put in re-education camps. Such can be easy to feel when only one moral perspective narrative is being constantly emotionally conditioned by one's mainstream media.

Lawyers firebombed a police vehicle during the George Floyd riots.

How did educated intellectuals lose control to a savage mob?

Were they rational? No.

Possessed.

The End of Moral Equivalence With Israel

As the count of murdered families, women, and children grows to over 1200 Jewish souls lost, with over 220 more still being held hostage, much of the world stunningly supports the Palestinians, or appeals to "both sides." But the question of whether there is moral equivalence between Israel and the Palestinians is simple to answer for any human being. The rest of the conflict is not much harder.

Some say one man's "terrorist" is another man's "freedom fighter", but the American colonists during the Revolutionary War never sent soldiers back to Great Britain to kill the families of the British Redcoats fighting in America. Humanity has spent centuries developing a moral code of military conduct from Hague to Geneva Conventions, to understandings and norms on our path toward being more civilized. If "terrorism" means anything, it is purposely targeting innocent civilians, without any military objective. Intentionally increasing the suffering of people you've never met, without even serving one's purported grievance, other than continuing the conflict for its own sake, is as evil as man can become.

Many in Europe are likewise slow to understand, as their warm welcoming European smiles are met with the horrors of London bombings, Madrid train bombings, Nice truck ramming, Paris attacks, Manchester bombing, London Bridge attack, Barcelona, Berlin Christmas market, and the list goes on. No country that enjoys peace and prosperity is immune anywhere in the world, though some cannot see they are next. History proves that even if we gave terrorists what they claim to want, they would increase terrorism anyway when they decide they want more. To side with Israel, the world merely need to look ahead at its own future.

Many raise as purported justification for Hamas' inhumane acts, the Israeli "occupation", "colonization", "genocide of the Palestinians", and "apartheid." Such highly-emotional trigger words conditioned into our moral consciousness from other points in history are now used to manipulate the world into anger and irrationality against Israel. The hope is to distract the world from mathematical reality, that Israel is only approximately eight thousand four hundred (8,400) square miles, in a Middle East that is 2.7 million (2,700,000) square miles. Meaning, 475,000,000 Arabs already control 99.6% of the land in the Middle East, and the Israel dispute is over the last remaining 0.4% for only 6 million Jews, in a tiny strip of land that is barely 10 miles wide at some points.

The math is important, because the Palestinians are Arabs. "Palestinians" are similar in language, culture, faith, identity, and cultural goals to other Arab Muslims like in Jordan, Lebanon, Syria, and Egypt, and can in 22 other Arab countries enjoy their food, culture, language throughout the 99.6% of the Middle East they already control. The dispute then, is actually between Arab hegemony, or the last remaining 0.4% for Jews needed for Jewish survival, proven by thousands of years of Jewish persecution across the globe.

We must also understand that the conflict is artificial. Iranian funding, and ideological power dynamics place the interests of Iran and the broader "Arab cause" above the Palestinians, who are prevented from acting in even their own interests. The Palestinian territories don't have elections, and when Gaza did, they voted in Hamas, who cancelled all future elections. Leadership accountability to the population is nonexistent. Hamas has never helped the Palestinians, but only Iran to the detriment of the Palestinians, and yet there is no other alternative. Six Arab countries have already made peace with Israel, and with Sunni Saudi talks currently ongoing in the disinterest of Shiite Iran, the terrorism and suffering of the mostly Sunni Palestinians is induced by outside agenda. Clearly, the recent Chinese-brokered Saudi-Iran peace deal signed this March needs some refinement.

The conflict is also fueled by ignorance of a history that only one side knows. The entire world recently renaming the ancient land of Judea in all our Bibles to something as uncreative as the "West Bank" proves it is intentional. If the name Judea sounds awfully Jewish, that's because it is, going back over 3000 years like the Jewish capital, Jerusalem.

The "occupation" is also a false narrative. There is no occupation in Gaza, as Israel withdrew completely under Prime Minister Ariel Sharon in 2005, and Hamas took over. Palestinians are free to travel south through Egypt anywhere they like, and Gaza has received significant economic investment. The problem is, when Israel relaxes border security it controls, Hamas brings in weapons to attack Israel with, as if enjoying forcing Israel to tighten restrictions further. Those who want more of Israel miss that this keeps happening every time they do.

But the question of Palestinian statehood is also a falsity, requiring we first ask some fundamental questions to determine whether they are a "people" and a "nation," because you cannot fabricate a nation from political pawns, or a dozen, separate, outside-funded terrorist groups.

When you ask "Palestinians" the meaning and origin of the name of their own people, they oddly stare like a deer in headlights. They have no idea other than believing they have simply always been called "Palestinians." But before 1967, they were simply "Arabs", and the "Palestinians" were actually the Jews living there. In fact, the Roman Emperor Hadrian named the land "Palestina" as an ongoing punishment against the Jews in the year 132 CE, after the Romans slaughtered half a million Jews. The Romans destroyed the Second Jewish Temple in the year 70 CE, said no Jews were to set foot in the land, and then re-named it after the Jews' ancient enemies from centuries prior, the "Philistines." The actual Philistines, who sailed over from Crete, were of course neither Arab nor Muslim. Palestinians actually say their name with an "f", "falastin", because there is no letter "P" sound in Arabic. The original word "Philistine" is not Arabic at all, but from a Hebrew word, for those who arrived from the Greek islands, whom the Hebrews called "Plishtim" "(פלישתים,)" which means "invaders". Not only does the Jewish word for invader prove the Jews were there first. One must ask, why would these Arab Muslim people want to name their would-be Arab Muslim nation after non-Arab, non-Muslim, Greek sailors, let alone based on the Hebrew word for them?

Terrorists make us angry so we fail to ask such rational questions.

And the important prerequisite questions continue. Do they have a unique culture and identity separate and distinct from other Arab Muslim cultures? Do they have completely independent interests, or are they intertwined with those of the larger Arab and Muslim world? Are they able to deter outside influence from other nations and ideologies? Do they have one set of common interests as a unified people? Are they able to pursue those interests as a unified people? Are they a unified people with a singular ideology that supersedes their divisions? Are they able to control their own terrorist factions? Do most of their leaders support separate individual groups and factions, or a single state ideology? And if so, is that goal to live in peace with Israel, or destroy Israel? Can they implement an education system that teaches acceptance of Israel? Do they have a leadership structure that speaks for all of them and their interests? Can they agree on and accept their own proposed borders? Sadly, the answer to these questions is "no."

The Palestinians have been betrayed by Hamas, and continue to support Hamas. They have been betrayed by Iran, and have been unable to sever their external influence. They have been unable to agree on what they want, negotiate in their own interests, or even counteroffer Israel's peace proposals. They have been betrayed by the Arab ideology that first waged war to destroy the Jewish state the day after its founding, and remain pawns of the same ideology today, to their own self-inflicted suffering. The adage, "a people get the government they deserve" is a principle of moral responsibility; not perfect, but the alternatives only get worse.

The Arabs who turned the Palestinians against Israel cause suffering because they actually turn against Islam. The "Naqba" or "disaster", when the Jewish state prevailed in 1948 over the Arab armies that attacked her, was because Arab ideology turned against the language of their own Quran, which explicitly gives the land of Israel to the Jewish people as does the Bible and Torah. As the Quran states, "And [remember] when Moses said to his people: 'O my people, call in remembrance the favour of God unto you, when he produced prophets among you, made you kings, and gave to you what He had not given to any other among the peoples. O my people, enter the Holy Land which God has assigned unto you, and turn not back ignominiously, for then will ye be overthrown, to your own ruin.'" [Qur'an 5:20-21] Arab anti-Semitic ideology supports neither the Palestinians nor the Quran, which states unequivocally, "And thereafter We [Allah] said to the Children of Israel: 'Dwell securely in the Promised Land" [Qur'an 17:104] All Islamic countries and groups need to do, is have the courage to follow the plain language of the Qur'an instead of those like Hamas who go against it. And that may be their best reason for releasing the hostages too.

Thus, it is clear that the dispute is not between religions, as all major religions' holy scriptures proclaim this land belongs to the Jews. The dispute is between the extremist ideologies that harm their own populations, and the rest of humanity. And the struggle, for the world to understand that we are all on the same side, against the terrorists, who are on nobody's side.

Humanity is in its Infancy in its Understanding of War, and it's About Time We Grow Up

Humanity's wars do not happen for the claimed excuses. If they did, we would be able to prevent them. Every causal reason given for every conflict by every expert during, and by historian after, is tangential, if not incorrect. Wars also don't happen for land nor money nor resources, nor over race nor religion. Despite that many Arabs want to destroy Israel because it is Jewish; these are all still workable challenges. We don't understand why wars happen at all, and we need to.

Humanity's wars occur because the world develops outstanding questions; moral, philosophical, definitional and others; and lacks the wisdom, logic, and mechanisms to properly address them in time. And the reason we cannot properly address these questions, is because humanity is not rational, but controlled by ideological collective hive minds which do not behave according to their claimed ideals, but separate, unconscious power dynamics.

To understand these dynamics would mean the beginning of the end of all war amongst mankind. This article is an introduction to some of these principles.

How are all mankind's wars the result of unanswered questions?

WWII occurred because of Europe's unanswered questions of what to do about Hitler's rise. The 2021 war in Ukraine occurred because the world didn't answer the question of how to satisfy Russia's security needs and NATO at the same time. The current war in Gaza is the result of world's outstanding questions about which side is morally right, how much land and support each side should have, and how Iran should be dealt with, how Gaza should be dealt with, as examples. Now the war continues because the world can't agree about what to do about Gaza, the Palestinians, the hostages, and Israel's future security.

Why cant mankind address such questions properly?

Humanity has difficulty answering the questions that would avoid war because moral perspectives diverge, appear relative, and humanity becomes irrational as we approach conflict. In 1939, the legitimacy of the nations of Europe rounding up Jews into ghettos, and the immorality of doing so, were simply divergent moral perspectives, each with their base of support. The world lost perspective of objective right and wrong, just like it has today.

Why do moral perspectives diverge?

It's not because of opposing interests, which can be negotiated, but a far deeper phenomenon. In the Lavool village in India in 2021, monkeys waged a species war against dogs, after a dog killed a baby monkey. The monkeys formed mobs, dragged every dog in the town to the tops of trees or buildings and threw them off, killing them. When the monkeys had killed every last dog in the village, they attacked a baby human child. Such hive mentalities have also gripped humanity when we were barely cave men with sticks and stones, with neighboring tribe of cave men attacking our camp, surrounding us, coming to take everything we had. That was war. Our brains reduced serotonin necessary for empathy and peace-building, the cerebral cortex was sidelined, and the primitive amygdala took control, sidelining rational thought.

In a similar phenomena, intelligent people involved in mob riots describe a strange sense of euphoria, drunk on the delicious cocktail of power, anonymity, and invincibility, like it was the most meaningful and fulfilling time of their lives, gathering to indulge in the same hive mentality, which is actually contagious group neurochemistry. The collective unconscious group hive mind was and is the primary driving force like Haidt's "Elephant and the Rider" analogy, and the individual cognitive mind a weak, distant secondary. We also see this in "group think" mentalities, when a group makes a less wise decision than any individual in the group would make, like the NATO "hive mind" decision to continue war against Russia through Ukraine.

How does misunderstanding such ideological dynamics lead to conflict?

The world failed to see Hitler's rise until too late because we misunderstood his ability to lead an ideology to war. The U.S. was wrong about Vietnam because we didn't understand how the Vietnamese ideology would unite with the communists. Russia was wrong about Afghanistan because they didn't understand why the Afghan Mujahideen ideology would unite with the U.S. Then the U.S. was wrong about Afghanistan because we didn't understand how the Taliban ideology was embedded in the Afghan collective psyche more than Jeffersonian democracy. The U.S. was wrong about Iraq because we didn't understand the Iranian leadership's ideology's ability to influence Iraq without a Sunni strongman like Saddam. And, Russia was wrong about invading Ukraine in 2021 because Putin didn't understand Ukraine became a separate ideology that would strengthen from opposing Russia. Now, Putin is wrong again about uniting Russia, Iran, China and North Korea, to try to gain advantage in Ukraine. A Middle East war is not in Russia's long term interest, as seen from the mob storming the Dagestan airport looking to attack Jews. And of course, China is also wrong in wanting to invade Taiwan, not understanding that most of Asia and more would unite against it.

The ancient Hebrew sages said all wars are the result of avoidable error. And John Steinbeck said, "All wars, a symptom of man's failures of thinking animal." Both were correct. As we are on the precipice of possible World War III, mankind is still in our infancy in understanding why we engage in war. We don't understand why wars happen, how to predict them, prevent them, nor avoid them. The United Nations is often unhelpful, and the peace-building landscape of our entire species, utterly primitive.

Individuals do not decide to go to war. But usually, neither do our leaders. No *person* does. Netanyahu didn't choose this Gaza war, but what if neither did any individual in Hamas? All of humanity's wars are caused by the ideological hive minds that control humanity, wired in primitive mankind since cavemen could wage war with sticks and stones. We become polarized over the outstanding questions. We are still those same cavemen. The more polarized we become, the more irrational, and the more we make mistakes. The ideologies gain power from emotionalized in-group versus out-group polarization, and break down the systems that have maintained peace, until eventually war is the only option left. Withdrawals from nuclear treaties like Russia is currently doing, pulling ambassadors, and cooperating with Iran's fanatical leadership, are primitive threats to tie our own hands behind our backs in escalations toward war because we don't know what else to do.

Why are so many treating Israel unfairly?

Why do millions in the West protest for Palestinians, but not the 350,000 of Syrians killed by Assad? Or the entire villages hit with poison gas? Or why not the 300,000 Kurds killed by Iran, or 4000 Kurdish villages destroyed by Turkey? Or the 1.5 million Muslim refugees now in 2023 being expelled by Pakistan back into the hands of the Taliban? Or why not the million Uyghurs held in camps by China or thousands of mosques they defaced or destroyed? Or what about the 1.2 million Armenians killed during the Turkish genocide for that matter?

The moral magnifying glass focus on Israel, or the West, is not rational, but because many, consciously or not, see Israel, or Western civilization, as the out-group. A similarly-polarized axis ideologically supports guilt-labeling Western countries for past colonization and slavery, but somehow not against Islamic slavery which lasted from the 7th to 20th centuries, nor Persians, Greeks or Romans for their slavery, nor Nigeria which had 2 million slaves until it was outlawed in 1936, nor Brazil which at one point had 4 million slaves, nor the slavery of the Turkish Ottoman Empire.

Most of the world does not choose its moral perspectives by rational deduction. It's ideological and tribal first, with even self-interest a distant second. Often, what we think is a rational problem is actually closer to a type of collective emotional imbalance that skews perspectives.

Why should all peace-loving nations support Israel?

Other nations have no moral right to criticize Israel for its response to October 7th, and usually do so out of hypocrisy. This is clear because in a hypothetical evenly matched bipolar conflict for world domination between Hamas and Israel, any rational self-interested person and nation would side with Israel. Those criticizing Israel do so merely because Hamas is not on their own border.

What is the underlying global moral question at the heart of the dispute?

The world should side with Israel because Israel is at the front line of moral questions that will envelop the entire world unless properly addressed. One such fundamental outstanding question being, whether it is moral to conquer them next. Israel will not, but Islamic extremists expressly intend to.

What America, Europe, Canada, and Australia didn't realize when they welcomed millions of Muslim refugees, is that while good individual Muslims may be grateful for the tangible benefits, that rationality is sidelined when dominant ideological group perspectives remain tribal. Leaderless ideologies naturally channel power to extreme personalities, and aggressively vying for power in non-Islamic lands provides inherent political gain. China and Qatar sent billions to U.S. Ivy League and over a hundred other universities, because they understand this tribalism better than idealistic intellectuals who pretend they can rationalize past it. While migrants, Palestinians, and Afghans can prosper better in democracies, the ideological compulsion to oppose out-group ideologies is often more powerful than their own self-interest. Concepts like "human rights" can be beautiful dreams or ideological war mechanisms depending on the ideological context.

Why did anti-Semitism increase 400% before Israel even responded to October 7th?

The attacks were so savage, burning babies and genital mutilation, that millions of ideologically like-minded felt compelled to take to the streets to connect with the intoxicating hive mind. Even leaders of nations feel compelled to join in. In September, 43% of Palestinians supported Hamas according to Palestinian Center for Policy and Survey Research, but a poll by the Arab World for Research and Development a few weeks ago found that 76% of Palestinians viewed Hamas favorably. As man gets closer to war, we become more tribal and primitive, and that includes more anti-Semitic. After October 7, many Western "liberals" woke up to see the true nature of the side they had unknowingly been supporting. It's an interesting question whether anti-Semitism rose because Hamas attacked, or, did Hamas attack because the global anti-Israel ideology became so strong, it destroyed peoples' rational ability to understand Israel is not committing genocide, apartheid, racism, or colonization, as so many mindlessly accuse. Apartheid was by European-origin settlers in South Africa against an indigenous population, but Jews are indigenous to Israel and have no other homeland. As "colonize" means people with one homeland exerting power over a foreign people in another land (ex: British colonies in America), only the Arabs can be colonizers, as Arabs are from Arabia, and Jews are indigenous to Judea. Jews by definition cannot colonize their only and home territory. However, this debate is not rational.

No matter how many parties in a conflict, it'll ultimately become bipolar, just like each of our world wars. This is why cheering the merits of "both sides", as the world tends to do, would have been unacceptable in WWII and every anti-Semitic persecution in history, only facilitating more conflict. The path to peace is not condemning both October 7 and Israel's response, but actually finding constructive solutions. The problem with criticizing Israel's policies is, unless one stands up for Israel's right to exist as a Jewish state first and foremost, they are helping the other side commit genocide while thinking they are merely being even-handed.

Why do ideologies with opposing values unite against Israel, the U.S., or the West?

Why do some self-described "liberals" support Hamas after they burned children alive? Why do some LGBTQ unite with Hamas who want to kill them? Why do some feminists unite with Islamic extremists who ban women's rights? Why do Islamic extremists unite with communists who ban religion? Why does BLM side with Palestinians and not Ethiopian Jews? Why does much of the world side with Islamic expansionists who openly plan to conquer them next? Terrorists are a scourge on all peoples, like pirates. Even if you make a deal with one, they just change their mind later, or another terrorist group attacks anyway. So why do they have so much global support?

The answer is, that ideologies do not behave per their claimed ideals, but unite even with opposing views for the combined power of doing so against another out-group. For example, while feminists should logically unite with the only Middle East nation that elected a woman prime minister, they unite with the Islamic extremists because their sub-ideologies unite against what they see as a common out-group in the establishment ideology of Western Civilization. Carl Jung said people don't have ideas, but rather, ideas possess individuals, and the development of this understanding is that ideas are living beings first, and will contradict their own purported values for the power gains. Friedrich Nietzche said, "In individuals, insanity is rare. But in groups, parties, nations and epochs, it is the rule." Both of them were also correct.

Why is winning the debate so difficult?

The logical points about the conflict, the "occupation", Palestinian statehood, and the rest, are secondary. The real problem, is the susceptibility of the Palestinian population to be ideologically polarized against Israel as the primary out-group. As long as that is the case, whether by Iran or Hamas or anyone else, it doesn't matter if you give Palestinians a state, education, and every one of them a Mercedes.

Ideologies behave like separate, living social organisms.

1. These separate living organisms have interests separate from the individuals and the leaders.

2. These separate living organisms have interests superior to the individuals and leaders.

3. These separate living organisms hide reality from and override the rational thinking of their adherents.

4. These separate organisms gain and channel power from in-group versus out-group polarization and conflict.

Mankind's wars can be viewed from the perspective of ideological entities that behave like separate living organisms as the invisible puppet-masters of conflicts. The measure of the degree to which an individual versus an ideology is the entity is called "entitativity."

Remember, when the Lavool monkeys attacked the baby human? It's because ideologies take on a life of their own separate from their values, and continue conflict like a living organism that wants to survive and grow. What else is suicide bombing, if not the ideology saying *it* is the entity, and sending the individual suicide bomber as the drone; not like a bee protecting the hive and queen, but harming the individuals for the power of the true entity, the hive mind. Clearly, what Hamas does is not in the interests of individual Palestinians.

We see such dynamics in the Communist Party in China which, the more it strengthens, the more it ironically sees the U.S. as a threat. We have it in NATO's mindless group-think mentality trying to destroy Russia in a decision no individual member would make. Of course, Russia and China, are both natural amicable power balancers with the U.S., and not Islamic extremists. Nonetheless, they now join those extremists in short-term-thinking partnerships in their own grave disinterest. They thoughtlessly empower terrorists, who would be more than happy to see Russia, China, and the United States in nuclear war destroy each other.

Why have we been so unsuccessful at destroying terrorist groups?

The U.S. has been unable to destroy the Taliban, and Israel has been unable to destroy Hamas and Hezbollah, because they are ideas. While Israel can and should kill every Hamas member in Gaza, it cannot destroy a terrorist group that is an idea without addressing the ideology, and its external support from Iran and Qatar. Remember with Amalek, Jews weren't told to kill just the fighters, but all *memory* of them. That's why the first thing Hamas wanted in exchange for the hostages was Netanyahu's assurance not to kill Hamas in Qatar. And why did Qatar broker the negotiations? Because they have been funding over $1.8 billion to Hamas.

Israel should defend itself regardless what the world thinks, as Golda Meir said. However, if Israel is truly wise, it will address not just the military challenge, but the ideological challenge of much of the world misguidedly uniting against it. Otherwise, what happened to Jewish minorities in so many nations in history that turned anti-Semitic will happen on a global scale, and that growing ideological entity will cause a world war, and a nuclear war, if not this time, soon.

How can we destroy Hamas and other terrorist groups that harm all sides?

After Israel kills every Hamas member it can find, here are some basics in dealing with terrorist groups and the harmful extremist ideologies that support them:

1. Extremists are like fish that can only swim in a sea of moderates. If you try to kill all fish in the sea, they'll just go into deeper waters and reproduce there. But the sub-group weakens when the tide, or broader ideology, recedes. Hamas is part of the broader Palestinian "cause", which is part of a broader Arab ideology, which is part of multiple broader Islamic ideologies which have both good and bad Muslims, pro and anti-Israel views.

2. Terrorist groups are an idea, and the only thing that can destroy an idea, is a better idea. Better options must be provided, even if the world

must push them upon a population. If there is no political debate amongst Palestinians, all out-group polarization is directed toward Israel.

3. It is sometimes better to divert powerful energy momentum instead of fighting it. "Free Palestine" should become "Free Palestine from Hamas". If "From the river to the sea…" became "From Jordan River to the Caspian sea, Palestine will be free", watch that phrase go extinct in days.

4. Empower moderate and constructive Islamic leaders to speak against terrorism, extremism, and Islamic conquest of Western Civilization, to help achieve balance and peace. An ideology's own leaders have greater ability to shape views. October 7th was in part response to power flow to moderate states like UAE, Bahrain, and the Saudis, versus Iran and Turkey's power gain from the Arab street's uproar.

5. Answer the outstanding questions and clarify moral positions. It is a perfectly moral position for non-Islamic countries to have strong relations with, do business with, and respect Islam in the Islamic world, yet still not want to be colonized in their own countries. Israel is at the forefront of a key moral question that shapes ideologies, which in turn shapes present and future conflicts. When the world accepts a better moral perspective than one that allows conquest of other cultures, political power will flow to leaders who seek balance. Until the world clarifies and unifies its moral position on this issue, we inadvertently unite those who seek peace and those who seek conquest, inviting expansion and conflict. A key moral question here is how to not be hateful, yet preserve indigenous cultures, a challenge India and many other nations wrestle with today, and all cultures may eventually.

6. Divide. Divide. Divide. Not with force, but ideologically. Not for conflict, but divide the elements of would-be unnecessary and irrational conflicts with truth and common sense as part of a broader peace-building strategy. Divide Palestinians from Hamas not acting in their interest. Divide all terrorists from their supporting populations, so peoples can have better opportunities. Divide terrorists from their leaders

and sponsors living in luxury. Divide Palestinians who want peace from those who do not. Divide Hamas from Iran. Divide Egypt and Jordan who refuse to accept refugees from the rest of the Islamic world that wants to help the Palestinians. Divide the Iranian leadership from the good Iranian people who have no interest in nuclear weapons or war. And divide Iran from its supporters, as Iran getting the bomb is not in Saudi, Russian, nor Chinese interests.

7. And finally, activate the prefrontal cortex.

You don't need an army to stop another army. Just ask them a question they can't answer.

In reality, neither Russia, China, nor the Islamic world necessarily need to be polar enemies, if we have proper balances in place. There is an entire universe of ways for humanity to build peace and prevent wars. We just need to find them, develop them, and implement them.

This is just an introduction to what comes next. Humanity is capable of such beautiful dreams, and such horrible nightmares. It's time we fight. Not each other, but for each other. It's time we made the world a much more beautiful place than it was. And so it begins.

The Trail of Mistakes to Nuclear War

With Putin desperate to weaken American support for Ukraine, he decided to form the closest military alliance with Iran the two nations have ever had. China's Xi thoughtlessly followed suit, believing anything that harms the U.S. is in his interest. Both these global strategic mistakes will make their partner Iran nearly sanction-proof in advancing its weapons capabilities.

Putin made the mistake of invading Ukraine, not realizing NATO would unite to oppose him. Biden made the mistake of unintentionally showing Putin a green light, taking no military action for months while Putin amassed 150,000 troops on the Ukrainian border. Biden not calling Putin throughout the length of the war was a mistake, as was having Boris Johnson reject Russia's peace offer just weeks into the war. Taking advantage of NATO's risky opportunity to try to weaken Russia was also a mistake. It's a mistake for Europe to continue buying energy from and selling equipment to Russia, but it's also probably a mistake for the U.S. to supply most of what Ukraine needs without conditions from Europe, such as at least not working against us. Of course, with policies that cause oil prices to rise, funding both sides of this war is also a mistake.

Such strategic global mistakes have consequences. With Russia and China going against their good judgment and allying with Iran, Iran predictably gained the opportunity it was looking for - a chance to distract the world with a terrorist attack causing a war in Gaza, so it can race toward the bomb. Not long ago, Biden Administration official Colin Kahl, then Under Secretary of Defense for Policy, said Iran would only take days to build a nuclear weapon. Iran is currently understood to be at over 84% enrichment, and will have weapons grade uranium at 90%, any day now.

Of course, Iran's ultimate goal is not just obtaining a bomb, but regional domination, and Shiite control of Islam's holiest sites of Mecca and Medina, which would cause the fall of the Saudi Royal Family. The Ayatollah said in 2015 that he wanted to "emancipate" Islam's holy sites from Saudi control, and now they have the world fooled and are apparently going for it. All one need to do is look at who benefits from this conflict. The Palestinians aren't benefiting, Hamas isn't benefitting, and Israel isn't benefitting. Iran is.

On December 10th, Netanyahu told Putin in a 50-minute long meeting that his cooperation with Iran is dangerous, and to cut ties with Iran.

But Putin is desperate to weaken U.S. support for Ukraine, wrongly assuming consequences he cannot foresee are better than those he can. Iran's leaders meanwhile, would be happy to cause a nuclear conflict that engulfs the U.S., Russia, and China.

All the Sunni Arab states made the mistake of falling for Iran's gambit. With a sweet spot for the Palestinians, much of the world fell for the distraction like a dog chasing a tennis ball. Turkey made the mistake of allying with Iran thinking it would gain an advantage regarding their mutual aim of weakening the Kurds, and it is also threatening to join the war in Gaza, like Algeria. While the Arabs are ready to fight tooth and nail over the Palestinians, their nations will ultimately fall under the control of the Ayatollah for their blunder. Egypt, Jordan, and other Arab states could have accepted Palestinian refugees to ease the conflict, but don't, to maximize it instead. In the United Nations, everyone's asleep, and simply focused on Israel and the Palestinians instead of thinking ahead. The riots in Europe have each country distracted, and they will all soon also be within Iranian nuclear missile range. Israel has been the only thing standing between not only Iran and Sunni Arab states, but Europe also. And the Palestinians, also mistaken, being Sunni, yet entrusting themselves to Shiite-backed Hamas.

Biden sent two aircraft carrier groups to address escalations, but apparently insisted that Israel will not expand the conflict in exchange for U.S. military support, preventing removal of Hamas leadership outside Gaza, and ensuring the conflict continues. The U.S. has also been working on freeing $6 billion and another $10 billion in frozen funds to Iran, potentially funding both sides of this war also. U.S. voters and media made a mistake also, not realizing how much Biden would run a continuation of Obama's disastrous Iran-empowering Middle East policy, and his feud with Putin. Biden's measured tit for tat responses also ensure that no party has anything to fear from a U.S. overreaction, and such intentional predictability ensures that this war also continues, and escalates. Making sure wars are endless appears to be an art form. Iran's proxy Hezbollah is looking to escalate, and the Houthis in Yemen are escalating from firing missiles and drones, to now a declared blockade of all ships to Israel.

Escalations will continue as long as there is gain from doing so, and little risk. Now, all Iran has to do, is increase the instability. The more chaos in the world, the more it benefits. Too bad the Iranian people didn't rise up and claim their independence on October 6th.

The Metaphysics of Anti-Semitism

It is the duty of every Jew to make the world a dwelling place for God, and be a light unto the nations. And that's true whether the other nations want that light or not.

Every generation, no matter how many times we say "never again", a plague called antisemitism causes peoples to rise to destroy us. What we do about antisemitism goes to the heart of who we are as Jews. If we know who we are, we know what we must do.

I had a dream once that everyone in the world loved the Jewish people, and there was peace on earth. They didn't just accept us, they loved us. And there wasn't peace despite us, but because of us. Once such a beautiful dream is born, we're halfway there.

The most Jewish response to antisemitism, is to take even the ugliest of humanity, and make something beautiful from it. But first, we must understand it. And perhaps by understanding it, all of humanity can understand itself better. Maybe that's the path to peace, not only with Israel, but worldwide. Let's see.

The IHRA's working definition of antisemitism "a certain perception of Jews, hatred of Jews…directed at Jews or their institutions…" is sadly insufficient. Even with some very good examples, it is incomplete. This definition doesn't help peoples understand the pivotal question: Why.

Even if every teacher walked into every university and high school tomorrow and told their class that the same plague had infected them that had infected the Nazis and every other generation that sought to grow their ideology by blaming and attempting to wipe out the Jews to resounding applause, they still wouldn't understand: Why.

Why do societies become anti-Jewish? Anti-Christian? Anti-American? Anti-Western, "Woke", "far-Leftist", and ultimately self-destructive?

Civilization after civilization has viewed Jewish commitment to our moral values - our loyalty to our moral code - as meaning that we can't be completely loyal to them, not realizing their civilization is built on such values. And as evil rises, any other civilization that likewise clings to a moral standard will face antisemitism-style resentment alongside us. Unless we figure out this conundrum, new ideologies with new "moralities" will continue to rise and seek to dominate the world, and necessarily attack Judaism, and then after Judaism, the broader Judeo-Christian world, American values, then all democracies, and then all other foundations of Western civilization, then all civilization, until they succeed in tearing it all down.

So since antisemitism is really everyone's problem, let's define Jew hatred, so we, and others, can finally understand it.

Here's my three-tier definition based on the Story of Esther, and its resulting political dynamics:

First part: Antisemitism is an irrational, emotional, subconscious discontentment about any aspect of human life that rationalizes itself under the fallacy of a complaint against Jews or their homeland of Israel.

Second part: Antisemitism spreads from a shared excuse among discontent individuals to societies where some use demagoguery for political gain focused on Jews or Israel as the scapegoated out-group.

Third part: And lastly, broader population segments unaware of these origins believe the false excuses of Jewish wrongdoing and are misled to a false moral perspective rather than properly answering outstanding moral questions.

In summary, antisemitism rises when new ideologies rise, the world loses moral clarity, and develops unanswered questions.

But to really understand, we must look at our whole story as the Jewish people. The story of our oppression goes back to when we were enslaved in Egypt nearly 4,000 years ago, continues through the world wanting to take Jerusalem from us and conquering it more than any other city in human history, and it follows us nearly everywhere we have gone, with few exceptions.

While our long story is so necessary I had to include it, I put it at the end as Appendix A, so you could skip to it, and the continue reading here.

Please see Appendix A (below) before continuing.

Most peoples don't know their full history, but we do. And now you know perhaps the hardest part of ours. And now you know why anti-Zionism is antisemitism. The world has a consistent pattern of trying to exterminate us; not just mobs but governments agreeing and cooperating with the mobs, rewriting laws against Jews again and again, until burning of our books, expulsions, and mass murder of Jews resulted. The Jewish people need the State of Israel to survive, and those who want to destroy us must first destroy the state of Israel. Thus, the necessarily-first target of many who hate Jews, is now Israel.

All hate is wrong, under any excuse. While many neighbors have had disputes, the hatred against Jews is unique. China and Japan, India and Pakistan, Vietnam and Cambodia, Turks and Armenians, Turks and Greeks, Turks and Kurds, Russian and Chechnyans, Irish and British, Ghana and Nigeria, Iran and Saudi Arabia, Saudi Arabia and Qatar: all at some point have made an out-group of their neighbor who is different. But that's not the same as antisemitism.

Excuses for hating the Jewish people have consistently been inconsistent, often contradictory. Too rich, too poor, too social, too isolated, …whatever the society deemed the most immoral conduct of that time, the Jews were accused of, as Rabbi Lord Sacks explained. If being the wrong religion was the worst immorality, then the Jews were. If opposition to government was, then treason was the accusation. If disease was the problem of the time, Jews were accused of spreading it.

The ultimate lesson from our history is that Jews haven't been oppressed for the claimed excuses, but for something inherent in much of human nature. The true reason, is below the surface, and unconscious.

We know Jew hatred is not rational because many new nations were formed since 1948, and their neighbors are not bent on the destruction of those new states. Hundreds of thousands of Muslims were killed in the Syrian civil war without major world protests. Turkey killed tens of thousands of Kurds with no major protests. Pakistan in 2023 expelled 1.5 million Muslim refugees, and there were no major protests. Likewise, a million Uyghur Muslims are held in camps in China, and China destroyed or altered thousands of mosques. Yet, none of these problems bring the public outrage we see against Israel. The UN has been averaging 15 resolutions a year condemning Israel, with perhaps one condemning Iran or North Korea. While none of these other countries are considered for termination, a recent Harvard CAPS-Harris poll said 51% of those polled ages 18-24 supported the end of Israel.

The intellectual are not immune. Modern, educated, cultured, intellectual Germans of the 1940's listened to Bach and Beethoven, and read Kant and Niztsche, while rationalizing away the extermination of six million of their neighbors. Without a grounding moral compass, Ivy League intellectuals today likewise self-rationalize their primitive anti-Jewish compulsions ever more articulately. The primal urge comes first, overrides reason, and the purported intellectual justification comes in hindsight. Antisemitism turns Ivy League doctors, and nurses, those sworn to protect life, into Jew haters. Those caught tearing down posters of kidnapped Israeli children cannot even explain why, unable to confront their own absent rationale. Universities like Harvard and Princeton are not merely antisemitic because the US blindly allows nations like China and Qatar to send them billions. 75% of Palestinians support the October 7th massacre, and many Ivy League professors do too. World antisemitism spiked by over 300%, not after Israel's response, but from October 7th itself, by those so enthused by attacks against Jews, they wanted to join in.

So antisemitism is a primal, unconscious, irrational urge inherent in mankind. And not only are the intellectual not immune, sadly they rationalize it away even better. We must understand that unconscious. Otherwise, as Carl Jung said, "unless you make the unconscious conscious, it will forever dominate your life and you will call it fate."

Humanity's conflicts occur because of unanswered questions.

Humanity's conflicts occur because of unanswered questions, typically about the correct moral perspective the world should hold. The world's moral perspectives are always changing, and every time new ideologies rise, so do our choices of moral perspective.

Such choices of moral perspectives are everywhere. As a simple example, if you are handing out ice cream to kids, do you give the largest portion to the biggest kid, the most well-behaved, or the one who is screaming for it the loudest? The basis for deciding is not just about satisfying the children then, but building a just society in the future. And if you didn't make a decision, the kids will fight over it.

The world's moral questions are likewise. The Russia/Ukraine war will continue until the security and territory questions of Russia and NATO are answered. China will want Taiwan until it understands why it shouldn't. And so on. Our international bodies and courts are imperfect, and often ill-equipped to decide such questions, thus resulting in conflict.

The most fundamental of such questions attack the Jews. When a newly-rising ideology seeks to replace the moral perspective of Western civilization, it necessarily must attack the Jews as the source of its foundation. When the Jews begin to appear immoral to the world, it's because the world faces a moral choice, and begins to lose perspective of what is right and wrong.

Without the Torah, morality would always change, causing conflict

Without the Torah, and Bible, morality would change every time new leaders and new ideologies rise. New, divergent ideologies must seek to destroy its mother to continue to grow. By definition, they must diverge or else they would be the same ideology. About 3,300 years ago, we stopped worshipping idols and began to believe in one God. Before that, new ideologies would form around newly-made up "gods", each having to fight to tear down the former civilization. Since the Jewish people accepted one true God, humanity advanced more than perhaps the million years prior. The Torah codified right and wrong, appropriate animal and grain offerings, because otherwise, ideologies would rise to offer a different value system.

Our Commandments documented "Thou shalt not steal" and "Thou shalt not covet", which seem obvious, but otherwise, new ideologies would arise to oppose the former. If a new ideology arose wanting to tax the rich 100% and declare it a higher morality, every poor person would be misled into following its leader, and we'd have a war. By the way, notice how communism is an economic system and should have nothing to do with religion, yet bans religion. It's because by honoring just these just two Commandments, communism can never take hold.

Even the way our domestic laws are interpreted changes. Twenty years ago, it was immoral, even if not illegal, to weaponize a legal system against a president with "excuse" crimes. The law didn't change; our moral perspective changed. And not only in America with the charges against Trump, but in Israel with the charges against Netanyahu, and in Britain with the charges against Boris Johnson. It's no coincidence that all leading Western democracies now all simultaneously have prosecutions by Left-leaning jurists against conservative, Right-leaning politicians. Either all conservative leaders in Western nations became criminals, or the moral perspective changed. As another example, Islam for many centuries preserved the Jews, but now an ideology within Islam opposes a Jewish state, because this rising ideology within Islam has a different moral perspective.

Unless resolved in time, the world's unanswered questions develop emotionally-polarized ideological divides between in-group and out-group, and cause conflicts.

In 2023, antisemitism is spreading again, now globally. As today's worst immoralities are racism, genocide, colonization, apartheid, war crimes, and ethnic cleansing, Jews are being accused of these, by many with no perspective. As it's not politically correct to attack Jews, the broader ideology attacks Israel, the only Jewish state and safe haven. The world lacks perspective, because these terms are ideological hammers to try to destroy Israel with, instead of values that can create a consistent framework for long-term peaceful coexistence. It's not that we don't all don't oppose racism, genocide, and colonization, but that any workable definition exculpates Israel. As I explained, "Apartheid" was by European-origin settlers in South Africa against an indigenous population, but Jews are indigenous to Israel and have no other homeland. As "colonize" means people with one homeland exerting power over a foreign people in another land (ex: British colonies in America), only the Arabs can be colonizers, as Arabs are from Arabia with the Islamic capitol in Mecca. Jews are indigenous to Judea and the Holy Land over 3,000 years, and Jews by definition cannot colonize their only and home territory. But without clarity, the world's changing ideologies are changing the very lens the world sees morality through.

All Jews ever wanted was to live in peace as Jews. Unlike Roman, Greek, Persian, Islamic, Mongol, and other empires, the ancient Hebrew civilization 1000 BCE to 70 CE is one of the only that in over 1,000 years there, never sent armies to colonize or conquer other distant lands. It's not in the Jewish culture. And, the Jewish population has always remained small, just a few million. Even waging war has strict requirements for the Jewish people, like seeking peace beforehand.

And of course, the entire Middle East conflict exists because Israel wishes to remain a Jewish state, called "Zionism". The word inherently has no negative connotation other than Jews wanting to have their indigenous homeland and culture. Of course, if someone is trying to annihilate the Jewish state and she defends herself, to the attackers, Zionism is a problem. Arabs who control 99.6% of the land in the Middle East, openly admit that if Israel's 0.4% were a 23rd Muslim state, there would be no problem. Somehow, the world's opposition to racism doesn't (yet) include this viewpoint, making the term "racism" devoid of meaning without this proper perspective.

So antisemitism is a primal, unconscious, irrational urge inherent in mankind. And not only are the intellectual not immune, sadly they rationalize it away even better. And it's based on diverging moral questions of the time.

Let's look deeper into resentment of the Jewish people.

History's antisemites reveal a subconscious feeling of lacking caused by Jewish contentment. The history in Appendix A shows a consistent pattern of not only expelling Jews, but forbidding Jewish practices and study, and even burning Jewish books. Jerusalem was attacked 52 times and conquered perhaps 20 times, more than any other people's capital in the world. However, it was never made into another empire's capital, as if done simply to take it away from the Jews rather than gain anything themselves. In Rome, the Triumph Arch of Titus depicts the image of Romans carrying away the Jews' Holy Menorah after sacking Jerusalem in 70 AD, celebrating taking Jewish possessions otherwise meaningless to Romans. Roman Emperor Hadrian renamed ancient Israel as "Syria Palestina", after not one but two of Israel's enemies, to try to keep Jews from returning to enjoy our own sacred homeland. What did a Roman emperor care if Jews returned later? As Mark Twain wrote in his letter of 1898, the Arabs left Jerusalem abandoned. The Arabs never rebuilt Jerusalem when they possessed it, and never made it the capital of any Arab state, not pre 1948, nor 1948-1967. They didn't want it until the Jews returned and rebuilt it. And now, there mere existence of a Jewish state is enraging, with that rage spreading like a virus to much of the world.

Why is there a historic pattern of an urge to deny us our faith, our Temple, our books, our rituals; appearing to be disturbed by Jewish happiness itself, and gain satisfaction from destroying it? Why does the world react to us, like a kid in a sand box, who would rather destroy our sand castle than enjoy building their own?

Dr. Jordan B. Peterson talked about an evolutionary hierarchy that goes back to lobsters over a hundred million years ago. In some ways humans still similarly decide dominance and leadership by primal indicators such as size and aggressiveness. However, if a group without these came to dominate hierarchies of competence as he puts it, it would be incredibly disturbing the those who believe they understand the natural order differently.

Why were Jews in Europe's history barred from many professions? Of course, there's no need to bar someone from a profession for being bad at it because their reputation would spread and they would be out of business. The only reason to bar a group from a profession is because they are too good at it. While Jews were less than one percent of Weimar Germany's population, they constituted 25 to 30 percent of the legal profession, including many leading lawyers, as Mark Twain wrote. Per his letter of 1898 "Concerning the Jews", he believed that nine-tenths of the hostility towards Jews in Russia, Austria, and Germany came from inability to compete successfully with Jews. Twain refers to a speech urging expulsion of Jews from Berlin, which confessed, "eighty-five per cent of the successful lawyers of Berlin were Jews, and that about the same percentage of the great and lucrative businesses of all sorts in Germany were in the hands of the Jewish race."

Jews have a culture of hard work, strong parental involvement, and focus on study, like many in some Asian, Indian, and other cultures, yet remain a minority everywhere except Israel. Even in the Middle East, Israel's 7 million Jews are a minority amongst 440 million Arabs. And the same way as minorities we faced antisemitism in communities, Israel is now facing similar as a small minority amongst other nations.

Twain's comment, of course, was before the modern state of Israel's reestablishment as a center of global technology and innovation known as the "start up nation", countless contributions to benefit the world, and having the highest per capita engineers and scientists of any nation. Jews have received 214 out of 965 Nobel prizes awarded between 1901 and 2023 for advances in areas like biology, chemistry, physics, and economics, discovering penicillin, and more. The Arabs outnumber the Jews 50 to 1, but have received only 3 Nobel prizes. According to a 1989 study of the greatest chess masters of all time, 28 out of the top 64 were Jewish, and so were a quarter to a third of top chess champions. Mark Twain wrote in 1898, long before the reforming of the modern state of Israel, "I am not the Sultan, and I am not objecting; but if that concentration of the cunningest brains in the world were going to be made in a free country (bar Scotland), I think it would be politic to stop it. It will not be well to let the race find out its strength. If the horses knew theirs, we should not ride any more."

In order to resent the Jews' success, the rest of the world must begin to covet, abandon personal responsibility, and take on a culture of blame rather than gratitude. Ideologies must change to see us as immoral, thus becoming evil themselves, because to adopt their moral perspective, they must go against the very laws of mankind, and any sustainable civilization.

But there is no need to fear or resent Jews. Let me explain some of the magic, whether one believe in God or not. Shabbat, the spiritual Sabbath day off from our material challenges, allows our subconscious minds to wander and return refreshed with new ideas. Since self-disciplined kids become more successful in life, we have Hanukah, when kids only one present a day for eight days. And we eat kosher, as the toxins in pork and bottom-feeding shellfish and other unhealthy animals may affect a healthy mind. The word for our people, Jew, comes from Yehuda, which comes from the ancient Hebrew word Hodaya, meaning gratitude or thankfulness. See Genesis 29:35 "*I will thank God and call my son Judah.*" The Jewish people, are literally the people of gratitude, which any spiritual advisor will tell you is the key to manifesting, because when you believe good things come to you, you notice opportunities better.

Resentment tears down civilizations, and you cannot have goodness and advancement without gratitude, an essential element of universal morality.

Jews are not "the chosen people" in any meaning of arrogance, but rather, we are "chosen" to bear a heavy burden through struggle, pain, and sacrifice for the benefit of all mankind. We are "a light unto the nations", whether the world wants that light or not. And when they don't, they try to put ours out. We have defined our existence, and whether other peoples have defined theirs or not, whether their morals are 1000 years old or invented yesterday, they have no moral right to hate us, and thus must turn immoral to do so.

The world would lose moral perspective without the Jewish people.

The world should celebrate and join Jewish accomplishments (and all others') as the potential of all mankind, and something to strive for. The other option, is to resent not just Jews, but all those who benefit and give meaning to our world. To resent those who taught us the value of human life, and gave us penicillin and a cure for Polio, is to turn evil. And such divergence in morality will find another excuse and cause conflict.

Abraham binding but not sacrificing Isaac taught humanity that human life is valuable, in a world where the value of human life did not exist as such, and people routinely sacrificed even children for the sake of countless undefined gods, hopes of better harvests and such. The Mayans and Aztecs sacrificed tens of thousands on their temples. Without the Jews, each one of us reading this, Jew or non, would likely be slaves if not a human sacrifice. And if the world ever destroyed the Jews, it would go back to such darkness.

The Ten Commandments; Thou Shalt Not Murder, Thou Shalt not Steal, Thou Shalt Not Covet, are prohibited precisely because they were not obvious to humanity, nor easy to follow. With Jews as the source of Western Civilization's moral conscience, we are the "canary in the coal mine" of evil growing. Any new ideology rising must distinguish itself from our moral core. The further they diverge, the more contrary to our foundational morals they become by definition.

Communism, for example, has to suppress all religion, as the entire system is based on the prohibited emotion of coveting what others have. When Jews warn against coveting, it is not just for our sake, but the coveters', for those in an ideology absent personal accountability will never be happy.

Nazism, as another example, rose against the Jews, because their ideology of white supremacy and killing all others couldn't not survive a challenge from Jewish moral values. Hitler admitted the battle for morality was between his and the Jews.

So my proposed definition of antisemitism helps societies look for the true causes.

"Antisemitism starts as an irrational, emotional, subconscious discontentment about any aspect of human life that rationalizes itself under the fallacy of a complaint against Jews or their homeland of Israel. Antisemitism spreads from a shared excuse among discontent individuals to societies where some use demagoguery for political gain focused on Jews as the scapegoated out-group. And lastly, broader population segments unaware of these origins believe the false excuses of Jewish wrongdoing and are misled to a false moral perspective rather than properly answering outstanding moral questions."

My definition describes and distinguishes between those who hate Jews due to their own personal deficiencies, those who facilitate antisemitism for political gain, and those well-intended who believe the propaganda of the first two groups. It is a way for antisemites to understand themselves. Because after all, it is not enough to cast blame, when they are the victims of their own mental illness.

Today, ideologies are rising and diverging again because of the world's outstanding moral questions.

Today's antisemitism started the day Israel was founded as a state on May 14, 1948, and five Arab armies attacked to try to wipe her out, and has existed since. They have been so insistent on their moral perspective, and so numerous with many of the nearly 1.8 billion Muslims worldwide and 57 Muslim states supportive, that this moral perspective has spread to much of the world. In 1800, the world's Muslim population was 91 million, in 1900 it was 200 million, and in 1977 there were 577 million Muslims in the world. By comparison there were an estimated 11 million Jews around 1900, 16 million before the Holocaust, and still less than 16 million today. The world Jewish population has been very steady, while the Muslim population has increased by 1.6 billion, now outnumbering the Jews now 100 to 1. By comparison, the population of the nation of Hungary was 7 million in 1900, and is now 9.3 million, growing only slightly.

With Islam's growing numbers, an ideology is rising among many in the previously-cohabitating and previously-intellectual Islamic world raising unanswered questions about the morality of Islamic expansion and conquest of not just Israel, but Europe, India, and every other nation, people, and culture. Good Muslims who seek cohabitation and balance are often sidelined amidst this surge. And the conflict for control of potentially-future Islamic lands, an inherent side effect of failure to address the world's outstanding related questions.

And no coincidence that the leaders of the ideologies against Israel today are the Russia, China, North Korea, and Iran, where there are likewise no free elections or due process. China even has a social credit and monitoring system giving it absolute control, and with their growth and expansion, are also becoming more aggressive than China has been historically, seeking to control the world's resources, ports, and even leaders.

Without the Jews, the world would descend into a darkness of North Korean oppression, Chinese surveillance and social credit systems, or an Islamist caliphate worse than the Taliban or Isis or Hamas, eventually consuming even moderate Islamic governments. Europe can't defend itself on its own, and America has lost much of its own moral compass. And thus, Israel is now the target of them all.

And that moral divergence is also behind the Palestinian antisemitism.

The world supported a Jewish state in 1947 with General Assembly resolution 181, acknowledging the Jewish return to our ancient homeland. But it probably wouldn't today if it had a vote, because the world's moral perspectives are diverging because of the outstanding moral questions of our time. Today, not only could the UN probably could not pass a resolution acknowledging Israel's 3,000 year indigenous history in the land, but it passed Security Council Resolution 2334 saying the 3,000 year old Jewish Temple does not belong to the Jews, destroying the UN's own credibility from its shortsightedness. On this path, is there any reason the UN cannot in the future declare away Buddhist, Hindu, Christian, African, or Sikh possession of its own beautiful cultural treasures likewise?

How did the world's moral perspective change? The world has never seen a civilization largely expelled from their indigenous homeland (although hundreds of thousands of Jews remained in the Holy Land), maintain their identity, and largely return and formally reestablish its nation after 2,000 years.

Israel should win the moral argument based on any reasonable moral perspective: Who was there first? The Jews. Who is indigenous? The Jews. Who has no other homeland? The Jews. Who better takes care of the land? The Jews. Who is the more unique culture? The Jews. Who needs the land to survive as a unique culture? The Jews. Who contributes more to the benefit of humanity? The Jews. Who made the land their capital? The Jews. Who is more advanced technologically, legally, intellectually etc.? The Jews. Who is more willing to live in peace with the other? The Jews.

So the world changes its moral focus to questions like: "Who is suffering more?" Or, "who has broader political clout?"

These are all various moral perspectives we can judge the parties through. However, in deciding which lens to see the Palestinian conflict through, here is the distinction: Some of these moral perspectives are applicable to cherish and preserve every other unique and beautiful nation and culture on earth, and the last two destroy them all. If Jews don't have an unconditional right to Israel, why would the French have a right to France, or the British to England, or Germans to Germany, Japanese to Japan, or Indians to India? Or for that matter, Ukrainians Ukraine, or Taiwanese Taiwan?

And if we don't all agree on that right, doesn't that necessitate challenge and invite conflict?

Simply, when choosing a moral compass, the world must choose one also applicable to them, for they are next. The question is a moral dilemma about group rights of indigenous peoples, and the answer is, if Israel cannot be a Jewish state, then no culture may have its own.

Once you understand the correct moral question, all the world needs is the courage to adhere to its own long term moral values necessary for peace.

During WWII, the world should have sided with the Jews more not because we were suffering more, but but because the Nazi ideology would conquer them next. Likewise, today, the world shouldn't side with the Palestinians because they are suffering more, but Israel, because a broader Islamic expansionist and conquest ideology unless moderated by a clear moral perspective will conquer them next. A clear moral stance from the world is needed to moderate and balance it.

And now that we understand the fundamental moral question, the sub-parts, and the rest of the conflict is easy. Every year on the Jewish holiday of Purim, we study the story of Esther, in which Haman had such a hole in his heart due to a Jew not bowing to him, that he was willing to pay valuable treasure just to kill Jews. And with this, the West can understand why Palestinians are willing to create their own self-induced suffering, just to try to harm the one Jewish safe haven. They didn't want two states before '48, or '67, and don't want two states now.

The missing fundamental understanding is that the Arabs who fight Israel aren't fighting Israel for the reasons most people in the world think, but simply because Israel is a nation of Jews they believe should not exist on what they believe is Arab land. A nation trying to forcibly conquer and take a land from another is bad enough. The world changing its morality to side with such a nation that can't even do the evil it desires to is globally destructive.

In simple words, in the world knowing whom to side with, Israel will never try to conquer the rest of the world, but an unmoderated Islamic world of 1.8 billion will.

If the world doesn't have this moral perspective, it can't answer the other diverging moral questions. For example: Is one Jewish life worth one Arab life? If the answer is yes, the conundrum then facing the world is, "are 15 million Arab lives worth 15 million Jewish?", because if yes also, given that option, that many would happily become suicide bombers to destroy the Jewish nation. Thus you have a moral paradox. You cant answer the secondary moral questions unless you answer the fundamental moral question first. And then, everything else is easy. Otherwise, a vote of 440 million million against 15 million in the UN is akin to 22 wolves and one lamb voting on what's for dinner.

Has the world decided the moral question, that the Palestinians, if they get a state, get a fixed and limited amount of territory, and only on the condition that they first prove they can live side by side in peace with Israel? No. If the world had unequivocally decided this and other such questions, the entire Arab world would be focused on teaching the Palestinians peace and coexistence with Israel. But without this moral clarity, those who want to destroy Israel are united ideologically, without a moral dividing line, with those who simply want to help Palestinians for humanitarian reasons. That's why so much of the world is against Israel. And Israel, left to defend herself for her survival. Thus, absence of moral clarity causes the conflict. So yes, the United States is correct for vetoing anti-Israel resolutions even though it may not well communicate why. And much of the world, sewing the seeds of its own later destruction.

As I discuss in my article, "The World is in its Infancy in Understanding War", as long as the Palestinians have a culture of resentment and coveting against their successful neighbor Israel, they will always be polarizable by any leader who takes a divisive position, no matter how bad that leadership is. Until the Palestinian culture meets the precondition of changing to one of acceptance of Israel, peace is metaphysically impossible, and a state would just be an increase in the same Palestinian momentum, and thus a launching pad for more conflict. This is not merely the reasoning for some Jews who do not want a Palestinian state, believing the prerequisite is impossible. This is a metaphysical prerequisite for peace.

How can you let Palestinians have an airport and open border, if they would just ship in weapons to launch at Israel?

How can you have a Palestinian state with a capitol in East Jerusalem, if they would simply let in perhaps 100 million Arabs to overrun Israel?

If someone reasonably believes in the "two state solution", underlying the question "why" is the assumption that the Palestinians are a separate and unique people who want to live in peace with Israel. You cannot have people with a moral perspective of blame and aggression toward their neighbor coexist with a more successful neighbor. In other words, a people of "gratitude" and a people of conquest cannot coexist unless the

Palestinian culture changes first. And it must change so much, that it accept less land than it could have had in '48 and '67.

Until then, Israel cannot morally be held to the impossible standard of being responsible for those willing to die to try to destroy her. And the world's lack of support for Israel without this precondition, only instigating war, against Israel, and soon themselves.

While the world should say the Palestinians should not get a state until they first prove they are able to live side by side in peace with Israel, they aren't able to say this. Neither self-inflicted suffering for political gain, nor terrorism for political gain, can be a permanent global moral perspective. If that were the world's moral perspective, it would consume every other nation. The world just doesn't realize it.

--

APPENDIX A

- We were enslaved in Egypt 1650BCE to 1450 BCE, and pursued even after we were set free.

- In 1400 BCE, we were attacked without cause by the Amaleks.

- After Kind David made Jerusalem our capitol in 1000 BCE, and Solomon's Temple was built in 950 BCE, our temple was destroyed in 733 BCE by the Babylonians, and we were expelled from Samaria in 733 BCE.

- We were held in captivity in Babylon 597 BCE, as described in the Book of Daniel.

- The Persians attempted our genocide in 475 BCE.

- We were sold into slavery across the Roman Empire in 70 BCE.

- In 63 BCE, 12,000 Jews died in Pompeii's conquest of the east.

- In 70 CE, the Romans destroyed the Second Temple; a million Jews died, and 97,000 were taken slaves.

- Thousands of us were killed in Cyprus and Egypt in 115-117 CE.

- We were forced expulsions from Judea 132-135 CE, which the Roman Emperor Hadrian renamed after our ancient enemy, the Philistines.

- There were forced conversions in Italy in 224.

- We were expelled from Carthage in 250, expelled from Alexandria in 415, expelled from Clement France in 554, expelled from Visigoth Spain in 612, expelled from Italy in 855 and all of France in 1181, as the hatred of us broadened geographically.

- There was rioting and attacks on Jews in Toledo Spain in 1212.

- By the Lateran council of Rome in 1215, Jews had to wear a badge of shame, pay extra taxes, and were denied public sector jobs.

- There were forced conversions in the kingdom of Leon in 1231.

- There was a ban on Jewish learning and Talmuds, thousands of Talmuds were burned in Paris France in 1239.

- We suffered confiscation of property, and were imprisoned, burned, converted and expelled in 1240 Austria.

- The council of Vienna declared Jews must wear pointed caps, and thousands were murdered in 1264 Germany.

- Jewish men, women, and children were imprisoned, and hundreds were hung in 1270 England.

- Jews were expelled and 180 were burned at the stake in 1276.

- The council of Offon denied Jews civic positions, and we had to wear a badge of shame in 1279 Hungary and Poland.

- Kind Edward issued an edict banishing all Jews from England, and many were drowned in 1290.

- Jewish refugees from England were also expelled from France in 1291.

- Libel of "desecrated host" and forced conversions of Jews took place in 150 Jewish communities 1298 Germany.

- We were expelled from upper Bavaria in 1442.

- We were expelled from Nassau 1478.

- We were expelled from Spain, and Calabria Italy in 1492.

- We were expelled from Portugal 1496-1497, expelled from Nuremberg 1499, and expelled from Naples in 1540.

- We were expelled by Pope Pius V from Papal states in 1569.

- We were expelled by Pope Clement VIII from Papal states in 1593.

- We were expelled from Frankfurt in 1614.

- We were expelled from Ukraine in 1649.

- We were expelled from American territories in Tennessee and Mississippi under Ulysses S. Grant in 1862.

- There were pogroms in Russia in 1880-1910.

- Then, the nations of the world weren't expelling us anymore. They wanted a "final solution", the Holocaust in Europe, in which 6,000,000 Jews murdered between 1933 and 1945. Many nations cooperated, and few nations helped, proving after thousands of years of this, that Jews need a place of refuge to call home to prevent our extermination.

- Despite 22 Arab countries controlling 99.6% of the land in the Middle East, the founding of the modern state of Israel on May 14, 1948 in our ancient Jewish homeland triggered immediate attack by the armies of Egypt, Transjordan (Jordan), Iraq, Syria, and Lebanon the very next day on May 15, 1948.

- Having failed in genocide, Egypt, Syria and Jordan attempted to attack and destroy Jewish state again in June, 1967.

- Then, again, when Egypt and Syria launched a surprise attack on the holiest Jewish day, Yom Kippur 1973, seeking to destroy the Jewish state.

- Meanwhile, millions of Jews, virtually all Jews, were forced to leave nearly all Arab lands in up to 22 nations, between 1948 to the present, leaving no Jews in the rest of the Arab world.

- Various attacks on the Jewish state by Iranian-sponsored groups surrounding Israel continued since.

- Finally, on October 7, 2023, some 1,200 Jews were murdered, and around 240 were taken prisoner.

Why Israel is NOT Violating International Law

A short, sharable explanation of why Israel is NOT ethnic cleansing, apartheid, colonizing, occupying, or racist. Of course, if Israel's enemies actually had these values and a debate about them, the conflict might soon end.

Common anti-Semitic tropes today are that Israel (or "Zionism" which is the desire for a Jewish state) is committing apartheid, colonialism, white supremacy, racism, ethnic cleansing, and genocide. These are false.

Of course, as Israel is not doing these things, it shows the accusers of these terms to likely be anti-Semitic in turn, or misled by anti-Semitic ideology. So let's define these terms the only way they make sense.

"Apartheid":

"Apartheid" is an Afrikaans word which means apartness, and only applies in South Africa. No definition outside South Africa has been agreed to. It was committed by a minority of European-origin Dutch settlers in South Africa against a majority indigenous population of native African Blacks. But Jews are indigenous to Israel, have no other homeland, and coexist with and give all ethnicities rights including to be in government, to be court judges, and to be in the IDF, so Israel cannot be committing Apartheid.

"Colonize":

"Colonize" means people with one homeland exerting power over a people in another distant land (ex: England controlling colonies in America). Otherwise it would be an "annexation." Jews are indigenous to the Holy Land with Jerusalem as the Jewish Capital for over 3000 years, after Moses led the Jewish people out of Egypt. Jews (from "Judea") who pray toward Jerusalem, have no other capitol or home country, and by definition cannot "colonize" their only and home territory. Unlike Roman, Greek, Persian, Islamic, Mongol, and other empires, the ancient

Hebrew civilization 1000 BCE to 70 CE is one of the only that in over 1000 years in the Holy Land (until many Jews were forcibly displaced) that never sent armies to colonize or conquer other distant lands. Only the Arabs which the "Palestinians" are can be colonizers in Israel, as "Arabs" came from "Arabia" in the 7th century, after the land had already been Jewish for 1700 years, with their Islamic capitol that they pray toward in Mecca, Saudi Arabia. In fact, the United Nations Declaration on the Rights of Indigenous Peoples General Assembly Resolution 61/295 of 2007, should, if the world is fair, apply to Israel as an indigenous Jewish people. It should protect Israel's rights including collective rights against the 1400 year-long colonization of indigenous Jewish land by other nations that have capitols elsewhere, including a lengthy colonization by Islamic peoples including to the Turk Ottomans, and eventually British Mandate. Israel being in the hands of the Jewish people is de-colonization, a respected value in the U.N. and everywhere else in the world. In fact, the Arabic name for Jerusalem is Al Quds, short for "Al Quds Bayt al-Maqdis", which means the place of the Holy Jewish Temple. In Hebrew, it is Beit Hamikdash, as Beit (בית) means house and mikdash (מִקְדָּשׁ) meaning sanctuary. Most of the Arabic language developed after Hebrew, in the 6th and 7th century. Even the Islamic Quran written 1400 years ago says that God has given this "Promised Land" to the "People of Moses" and "Children of Israel"? (Qur'an 5:20-21, and 17:104)

"White supremacist" or "Racist:

"Racism" is hatred toward another race. The U.N. passed a resolution saying that "Zionism is a form of racism and racial discrimination" in 1975 pushed by the Arab League, and many Muslim majority countries, but revoked it in Resolution 46/86 in 1991. Firstly, almost half of Jews are Saphardi, Mitzrahi, South American, or African, with Israel as their only home, so Israel cannot be White supremacist or racist as Israeli Jews are of every skin color, nationality, and origin. Second, Israel has 2 million Arabs living within its borders, Druze, Armenians, Samaritans, Bahai, and other cultural and ethnic groups. The Arabs in Gaza on the other hand are 99% Arab Muslim and allow virtually no other ethnicities or religions. Palestinians also admit that if Israelis converted to Islam, there would be no dispute, and the dispute only exists because Israel is controlled by Jews. Finally, there are many nations that bar or restrict certain religions or ethnicities, their holding of office, or have other restrictions to preserve an ethnic or cultural identity. Some nations ban all religion except Islam; other nations effectively ban Islam, just as examples. Israel has no such restrictions, but is blamed far more than countries that do. Islam also has a rule that other religions are beneath them and have to pay Muslims a tax called Jizya. Thus, if anyone, the Palestinian Arabs are racist and supremacist, but Israel is not.

"Genocide":

Per the 1948 Convention on the Prevention and Punishment of the Crime of Genocide, "Genocide" is intending to destroy, in whole or in part, a national, ethnical, racial or religious group. The "Convention on the Prevention and Punishment of the Crime of Genocide" (1948) was written in the wake of the Holocaust, because Hitler hated Jews so much, he declared he would and proceeded to exterminate the entire Jewish race, when he had other, more humane options, like expulsion. There are 440 million Arabs in the Arab world, and only 7 million Jews in Israel. There are 22 other Arab states, and no other Jewish state. Israel only uses force against Palestinian Arabs to stop, prevent, and deter acts by those who admittedly intend to destroy the Jewish state "from the river to the sea", so only Arabs can be committing Genocide. Five Arab armies tried to destroy the Jewish state the day after its founding on May 14, 1948, and again in 1967, and 1972. Iran threatens to destroy the Jewish state as does Hamas' Charter. Assad killed 350,000 Syrians and bombed villages with poison gas, Iran killed 300,000 Kurds, Turkey destroyed 4000 Kurdish villages, Russia invaded Ukraine in 2021 to destroy the entire existence of the nation of Ukraine, and the world is not aggressively accusing any of them of genocide. So, Hamas-led Palestinians who are Arab, and Iran's leadership, are trying to commit genocide against Jews, but Israel is not and could never destroy a population of Arabs 50 or 100 times its size.

"Aggressor":

As an "aggressor" is a country that attacks first, Israel is not an aggressor, as it has only fought defensive wars after repeated attacks by Arab groups and states trying to destroy her. Every nation has the right of self-defense. The "right of conquest" is the historic international law principle that a nation that possesses land after force of arms has ownership. While wars of aggression are illegal under the U.N. Charter, land acquired in a defensive war, as Israel has reacquired its ancient land after being attacked, is legally owned by Israel.

"Occupation":

The Hague Convention of 1907 defines "occupation" in Article 42 "military authority over the territory of the hostile state", and as a treaty between nations, it applies, per Article 1 only to signatory nation states, "Article 1. The Contracting Powers shall…" Palestinian territories are not a nation state and not a treaty signatory, so Israel cannot legally be an occupier over the territory of a non-state and non-signatory. The term is intended for an existing nation's land being occupied, not an unincorporated territory, because a nation can be signatory to agreements and negotiate, whereas a terrorist group has no accountability and cannot. From 1948 to 1967, Jordan controlled the "West Bank" which is really Judea and Samaria, and Egypt controlled Gaza, and they were not accused of "occupation." Israel ended occupation of Gaza in 2005 withdrawing every last troop and civilian. Instead of Gaza forming a state, they launched near-daily missile attacks culminating in the October 7th attack killing over 1200 civilians. Thus, Israel's "security presence" is made necessary by her right of self-defense.

"Ethnic cleansing":

While "ethnic cleansing" is not a defined crime under international law, a U.N. Commission described ethnic cleansing as S/1994/674 as "… *a purposeful policy designed by one ethnic or religious group to remove by violent and terror-inspiring means the civilian population of another ethnic or religious group from certain geographic areas.*" Israel is not ethnic cleansing, and would not be even if Palestinians were displaced. Many Palestinians want to leave Gaza voluntarily. But even if not, Israel is not selecting people for displacement based on ethnicity or religion, but being part of a terrorist group and its supporting population Israel is defending herself from. Palestinians are 99% Arab and Muslim and are homogenous because they are racist towards other ethnicities, but that's not Israel's fault. Israel allows millions of Arabs and Christians and other groups within her borders. The people of Gaza do not. So, if a group that is homogenous because their own racism is displaced, their racism does not transfer onto Israel. Pakistan expelled 1.5 million Afghan refugees from their country in 2023, and a million Uyghurs are separated and held in camps by China, and none of their leaders are being aggressively accused of ethnic cleansing.

So, since Israel is not doing these things, no person, organization, or body, or even the United Nations, should be wrongfully directing these accusations against Israel, in speech or in resolution. It harms both these values, and the credibility of their people and institutions. People who seek constructive discussion of Israeli policy should be aware that they may be unintentionally aligned with those who don't stand up for the absolute right of Israel to exist as a Jewish state, but who would distort these values to attack and try to destroy Israel as a Jewish state. Such mislabeling of Israel spreads hatred of Jews, as anti-Semites have done throughout history. And anti-Semites were never the good guys.

Writings on the Rafah Border with Egypt

The Three-Option Plan for Exploring Simultaneous Palestinian Statehood and Relocation

The recent phone calls between President Biden and Prime Minister Netanyahu after the Prime Minister's statements opposing a two-state solution present the need for a new idea. Israel's approach, and the entire world's approach to the Palestinian problem needs a new perspective.

Israeli officials have effectively said all of the following in recent weeks:

- that Palestinians should be relocated,
- that Israel has no intention of relocating them,
- that Israel must control security from Jordan to the sea,
- that there will be no Two-State solution, and,
- that a Two-State solution is not completely off the table.

Even amongst Israel's allies, frustration with Israel is growing, and support weakening, after Netanyahu's comments. Still, Israel cannot be expected to shoulder the burden alone of a global problem instigated by outside forces.

And meanwhile, Iran is taking full advantage in instigating more conflict while expanding its nuclear program.

As reported by the Jerusalem Post, President Isaac Herzog recently said in Davos, "If you ask an average Israeli now about his or her mental state, nobody in his right mind is willing now to think about what will be the solution of the peace agreements." And that Israelis have, "lost trust in the peace processes because they see that terror is glorified by our neighbors." Most Palestinians don't want a two-state solution either, only 17% according to a November 14th Survey by the Arab World for Research and Development, with 74% wanting a Palestine "from the river to the sea".

The world is likewise polarizing around this conflict, as the international community doesn't understand that any Palestinian state under the current Palestinian ideology would just be a means to "bring in the heavy artillery" - a stepping stone to better attack Israel from. A Palestinian capitol in East Jerusalem or militarized state would just be an excuse to bring in a million soldiers and start an all out war against Israel that would draw the whole world in. The past 16 years in Gaza is proof of what the Palestinians would do with full statehood. Yet, the status quo cannot continue forever either.

Netanyahu is under increased international pressure to accept a pathway to a Palestinian state, by an international community who do not understand or care that most Palestinians' goal is to destroy Israel, not live alongside her in peace. Meanwhile, the Jewish people are a moral people, who neither want to control Palestinians nor accept their terrorism.

So, here's one possible solution. Israel need not choose between Palestinian statehood or relocation; it can explore both options simultaneously. If the Palestinians continue to oppose Israel, under such a plan, they will increasingly create a global momentum for their relocation into the Sinai. At some point, any idea will be better than continuous war with no plan whatsoever. Especially, as allies like Saudi Arabia, regardless reality, demand some path forward to be able to unite with Israel to focus on Iran's imminent nuclearization.

A new perspective of the problem:

Humanity approaches the potential resolution of all wars incorrectly, per theories I discuss in my article, "Humanity is in its Infancy in Understanding War." The problem is not so much individuals nor their leaders, but the ideological dynamics of the group collective psyche, which controls both, and is the true reason for all war. Since the dawn of mankind, humans are tribal, and face threats in groups on a subconscious, primal level. These collective group "hive" mindsets are the primary motivators of human group conflict, and rational analysis of individuals a distant secondary. Abbas and Arafat were unable to make peace as adherents controlled by the broader ideological entity, the "Palestinian cause." While often tribally-defined in the region, groups can actually form around nearly any shared interest or ideology.

The Palestinian problem is not ultimately about land, nor money, nor religion, though many Palestinians hate Israel because it is Jewish. The problem ultimately is the dynamics of the group's ideological in-group versus out-group polarization, which I explain in my writings is the key to all war. The problem is not one of making peace with leaders, but one of ideological momentum, and must be dealt with on a metaphysical level, according to principles I am developing within the PeaceMatrix™ system called PeaceMatrix™ Entitativity theory. These principles resolve conflicts by strategically addressing the ideological dynamics of the parties' collective mindsets.

A key cause of the current paradigm is that today those who merely want to alleviate Palestinian suffering are ideologically aligned with those who want to destroy Israel. They have the same initial goal - a Palestinian state. This problem exists across the Palestinian territories, the broader Arab community, the broader Muslim community, and the even broader international global community. Without a political process, the Palestinians have nowhere to direct their inherent, tribal, ideological in-group versus out-group polarization except at Israel. They can't be constructive in improving their own lives when their focus is opposing Israel. Anyone who attacks Israel whether verbally or militarily, like Hamas, is elevated within Palestinian society, because the broader ideological collective psyche's dynamics channel status, power, resources, and benefits to those who instigate opposition to Israel.

The pressure from this global problem is wrongfully laid on Israel's shoulders; not Iran for sponsoring terror, not Arab states for not helping more, and not the broader international community for accepting terrorism and making unreasonable demands. The international community can no more demand a Palestinian capitol in East Jerusalem than demand that India make half of New Delhi a capitol of a new

Islamic state; or half of London, or half of Paris, or half of Moscow, or half of Beijing be capitols of new Islamic states in those countries. Since a main reason for the current paradigm is that those who genuinely want to help the Palestinians are united ideologically with those who want to destroy Israel, the first thing we need to do is separate those two viewpoints and groups. The strategy must help one at the expense of the other.

A second reason for the current situation is that with every peace offer in the past 75 years, Palestinians have felt they had nothing to lose, and believed they had only to gain from continuing the conflict. It's "try to conquer Israel", or, "stay at the status quo". Like a mugger to whom you must return his weapon after you thwart his attack, he has nothing to lose from trying again. Culturally, there is no equivalent word in Arabic for the concept of "compromise," only "resistance", and "more", whether they have 1% of the land, or 99.6%, which Arabs already control of the Middle East anyway. Approximately 72% of Palestinian support October 7th style terrorism, and the vast majority support never-ending war against Israel until Israel no longer exists. Gaza has been its own de facto independent Palestinian state for 16 years and all it has done is wage war against Israel. Some say it is still less moral to keep the Palestinians in that predicament than to relocate them, not just for Israel, but for other Arab states not taking Palestinian refugees also. We must therefore also utilize a carrot and stick approach, that provides some consequence, alternative route, or both, if Palestinians reject the offer or fail to meet reasonable requirements.

A third understanding is that the Palestinian "cause" is a living ideological entity, and if peace were made, the Palestinian cause would die, and it doesn't want to die. See more on PeaceMatrix™ Entitativity theory in my other writings. So, the third thing we must do, is address the fact that the Palestinian cause ideology wants to continue, in some form.

The solution:

The solution should restructure ideological polarities in favor of Israel, peace, and stability, and helping Palestinians at the same time. The solution is to change the polarization from Arab cause versus Israel, to, those who want peace versus those who want a different peace. And, to do so within a controlled framework. In other words, let's exchange an unhealthy polarization for a healthier and more constructive one. The plan presented here for examination is an example of how we might look at potential solutions and draw those lines.

The Plan:

Propose a working group for the simultaneous exploration of two options; either a Palestinian state based in Ramallah under preconditions, or in the Sinai.

Such a proposal would be rejected outright, but it doesn't matter. Many Palestinians will always want to destroy Israel, but it doesn't matter. The plan is designed to regardless have countless benefits, and ultimately may be self-implementing. Here's why:

The plan is a 3-part plan that if Palestinians meet a set of Israel's prerequisites, then a demilitarized Palestinian state can be further explored with a capitol in Ramallah; but if they do not, then the Palestinians are relocated, either to a Palestinian state in the Sinai if Egypt allows, or absorbed into other Arab societies if Egypt does not.

The solution is not the plan itself, nor its acceptance or rejection; the solution is the reasoning behind the plan, as explained below.

Option 1)

The Palestinians have a timeframe to meet a list of prerequisites ensuring Israel's security, which if they meet, with the world's help, will move the parties closer to a demilitarized Palestinian state after that time. Israel's list may include:

- Destroy Hamas
- Deradicalization the ideology of Palestinian society
- Accept Israel as a permanent Jewish state with Jerusalem as its Capitol
- End violence and intent to destroy Israel
- Palestinians must reject all external anti-Israel influence, including Iranian
- The international community will sends monitors to de-radicalize education, promote peace education, control political process to allow diverse candidates to run, candidates must hold elections yearly, and must not promote jihad, hatred, and opposition to Israel
- The international community will help install democratic institutions, constructive cultural, sports, and community activities and bring economic incentives
- International community will stop all Iranian and external anti-Israel influence and nuclearization
- Palestinian territories are demilitarized and Israel maintains control of security from the Jordan river to the sea
- Palestinians must accept permanent borders and acknowledge that further attacks against Israel by anyone including Iranian proxies will cost them land
- Palestinians who want to relocate are allowed to move to other Arab countries temporarily or permanently to ease pressure off conflict
- Israeli and other leaders have an ongoing public debate about all options. Ultimately, Israel is right, and the more open and constructive debate, the better.

Option 2) and 3)

If Palestinians do not meet these prerequisites, despite it not being Israel's or the world's first choice, many Palestinians may end up being relocated for their own good to:

Option 2) a new Palestinian state in part of the Sinai if Egypt allows, or,

Option 3) absorbed into other Arab states if Egypt does not. The choice between #2 and #3 being placed in the hands of Egypt, other fellow Arab states, and the broader global community.

The plan is a compromise between the world wanting an impossible two-state "solution" by which the Palestinians would continue an enhanced permanent conflict, and those who believe relocation is the most moral remaining option after Palestinians have rejected peace offers for 75 years. Again, the question is not whether relocation is moral, but whether it is more or less moral than allowing them to suffer in endless conflict.

Benefits:

Exploration of these options can be a unilateral process. Even if opposed by all parties, the international community will be presented with another option for Palestinian statehood which is better than no option, taking pressure off Israel.

The reason many Israelis oppose a Palestinian state is because they believe Option #1 is impossible to achieve. Fine. This plan will motivate the international community to try harder at Option #1, while simultaneously building momentum for Options #2 and #3, in case #1 fails.

Those who oppose relocation have another option - to try harder to accomplish Option #1 above. Arab states who oppose taking in Palestinian refugees can support Option #1 or #2. If Egypt opposes an Option #2, it can support Option #1 or #3.

Think of ideological energy as water. Instead of all the flow being directed at Israel, it is redirected three ways, which will compete against one-another to offer Palestinians the best possible life.

Ultimately, the longer the Palestinians refuse to accept Israel, and the more Option #1 proves impossible if that is the case, the more attractive relocation as Option #2 or #3 becomes to the world.

Explanation:

The plan motivates Palestinians and those globally who want a peaceful Palestinian state alongside Israel. It also lays parameters, without which, Palestinians' ability and desire to attack Israel would continue the conflict even more if a state were established. This process is essentially a test, to see if Palestinians are able to accept Israel and live in peace, or only want a state to better attack Israel from.

Many in the world who want a Palestinian state do not care whether it is next to Israel, or in the Sinai. This plan is reasonable to those who want to help the Palestinians but don't want to destroy Israel. Thus it creates a debate between them. It draws an ideological polarizing line as the debate grows, between those who just want a Palestinian state versus those who want to destroy Israel, instead of everyone ideologically united against Israel. When one option becomes more difficult, momentum in the form of ideas, resources, and policies will flow to the other options. Cyclically, momentum will grow for all three options as their proponents oppose each other and advance the debate, as human nature requires.

If the Palestinians reject a demilitarized state peacefully alongside Israel, and prove they only want a state as a stepping stone to destroy Israel, the international community will increasingly see relocation as justified. After all, if Palestinians don't want peace, it is immoral for the whole world to force them to stay there.

This plan answers the component missing from every other peace proposal, which is, what happens if the Palestinians reject it.

Palestinian society will engage in debate polarized between those seeking to destroy Israel versus those seeking a peaceful state with self-governance. It bifurcates the "Palestinian cause" into its two primary subcomponents.

Palestinian society will also engage in debate polarized between those seeking a limited peaceful state alongside Israel, versus those seeking a larger and more lucrative state in the Sinai, which could have substantial global investment as a major international center.

Palestinian society will also engage in debate polarized between those seeking a state versus those seeking to relocate to other Arab states, as international pressure for Arab states to accept refugees grows. Thus, it also separates the Palestinian "cause" from individual Palestinians' interests also, as nothing has caused as much suffering for individual Palestinians as the "Palestinian cause."

The international community that wants Option #1 will have incentive to put pressure on Iran, as the plan also separates Iran's goal of destroying Israel. The "Palestinian cause" is a very powerful ideology, and Israel can utilize that cause and turn that energy back against Iran (or even Russia and China for supporting Iran) for causing the Palestinian suffering by supporting Hamas for Iran's own political gain.

Explained another way; if Israel displaces the Palestinians, Israel faces 100% of the global pressure. If Israel tells the world Iran is the problem, the world won't listen, because it's easier to just put pressure on Israel. But if Israel says it doesn't want to but may have to displace Palestinians unless Iran stops terrorism and nuclearization, and if Israel offers a potential solution if Iran stops, then Israel has created a polarity, and Iran faces increased pressure.

As the Palestinians continue demanding a state but fail to meet the reasonable prerequisites, the debate will increase regarding a Palestinian state in the Sinai.

Of course, Egypt will vehemently oppose, but as Palestinians fail to meet Option #1, the debate will increase around Egypt and Arab states taking refugees, Option #2 and Option #3, disbursing pressure that is currently united 100% against Israel. The Arab states are added to the negotiating table, as they would rather have a Palestinian state in the Sinai than take in refugees, and Egypt vice versa. Egypt is over 1,000,000 square kilometers, with over 2,450 kilometers of coastline. It is 45 times the size of Israel at only 22,000 square kilometers. The Sinai alone is 60,000 square kilometers or about three times the size of all of Israel. Not only would the Sinai make more sense for Palestinians whose aim is not destruction of Israel, but the economic opportunities would be massive for the entire region.

The media, experts, and world leaders will vehemently discuss all the Arab opposition. But after they've covered every conceivable story about why it won't work, some might start asking how it might. Some might ask, "why should countries take in suffering Palestinian refugees?" Or, "How can countries support a better life for the Palestinians while also protecting Israel?"

Eventually, Palestinians will realize that unless they get their act together, they will be displaced. Egypt will realize that unless it takes some Palestinians, it may get stuck with all of them. The Saudis will realize that unless they work with Israel now, Mecca and Medina will soon be controlled by Iran. And poor Arab countries will realize, that the West will pay billions to take in some Palestinian refugees. And so on.

And the world will realize the value of asking not binary, but open-ended constructive questions:

What are the benefits of a Palestinian state in the Sinai?

What are the benefits of relocation for the world?

If Israel must eventually relocate the Palestinians, Israel will have given the Palestinians yet another option for peaceful statehood first. The plan will allow the world to better accept relocation, and for political channels to form around relocation as a natural product of the pressures of the conflict.

If Palestinians continue to fail to meet the prerequisites, relocating them becomes an increasingly-attractive solution to much of the world because they can still have a state, that they claim they want, just in the Sinai, where they are not increasingly conquering Israel. The plan takes the pressure off squarely Israel and puts it on three different groups, drawing ideological lines between those who want a Palestinian state in the Sinai and those who do not, and between those who want Arab states to take in Palestinian refugees and those who do not. If the Palestinian problem continues, it is the Palestinians' fault for wanting to destroy Israel, AND Egypt's fault for not allowing a state in the Sinai, AND the other Arab states' fault for refusing take in refugees.

The plan unites those wanting a Palestinian state in the Sinai with Israelis who want to transfer Palestinians.

The plan can also be based on aspects of Trump's Deal of the Century.

The plan fuels increased global debate as many believe Arab states should shoulder their burden and take in Palestinian refugees. If keeping Palestinians in an "open-air prison" is immoral, and if crimes against Israel are alleged, then all Arab and all other global states who refuse to accept Palestinian refugees are complicit. This debate also parallels a broader debate about African, Arab, Asian and South American countries shouldering more of a regional refugee burden rather than mass migration to other continents where cultural differences make integration difficult.

The plan fuels debate between Muslims who want to conquer Israel, and those who believe in following the literal Qur'an which says Israel belongs to the Jewish people, empowering the latter. As the Qur'an states, "Remember when Moses said to his people. 'My people, remember Allah's favour upon you when He raised Prophets amongst you and appointed you rulers, and granted to you what He had not granted to anyone else in the world." (Qur'an 5:20). "My people! Enter the holy land which Allah has ordained for you; and do not turn back for then you will turn about losers." (Qur'an 5:21) "And thereafter We

[Allah] said to the Children of Israel: 'Dwell securely in the Promised Land" (Qur'an 17:104).

Maybe the Nakba and 75 years of suffering is because Muslims went against the language of the Qur'an, and maybe following the scripture and giving up trying to destroy Israel and letting Israel "dwell securely in the promised land" will yield better results. Maybe it'll be better for all Arabs.

It'll change the current dynamic from increasing pressure on Israel the longer there is no Palestinian state, to increased global acceptance for relocation of Palestinians the longer Palestinians continue their conflict against Israel. It'll give the Palestinians a better chance at a state if they are peaceful, a chance at another state if they are not, and a better chance at better lives regardless. It'll place the choice between a peaceful state and relocation in the hands of the Palestinians. If the Palestinians can act in their own independent best interest and separate themselves from an external genocidal "cause" against Israel's existence, they may have a state, and if not, they themselves may be better off elsewhere.

If Arab states have a path forward, they can realign with Israel to address Iran's nuclearization, and they and the whole world will be better off than being in a never-ending stalemate that increasingly threatens to draw the whole world into conflict.

By decreasing pressure on Israel, decreasing the benefits from continuing conflict and the status quo, and exploring different options simultaneously, all parties will have a greater stake in exploring solutions more constructively.

SUMMARY: The Palestinian problem is not about land or resources but an ideological conflict; and based on Daniel Ben Abraham's new theories on Entitativity, the conflict can be resolved by correctly addressing the ideological dynamics.

Entitativity analysis: Palestinian relocation

A conference of Israeli Ministers and Parliamentarians in Jerusalem on January 28th 2024 examined the idea of voluntary relocation of Palestinians. In attendance were 15 members of Knesset and 11 Parliament members, including National Security Minister Itamar Ben-Gvir, Bezazel Smotrich, and over a thousand others, discussing related ideas.

But to gain increased global support, the world must understand why a relocation option may be not only advantageous, but a necessary component of even the most liberal approach to the peace process. As I discussed in my article on my Three-Option Plan, Israel is only keeping Palestinians out of Israel; but it is the rest of the world's mindless policy of not accepting Palestinian refugees that is keeping them in an "open-air prison." Not only is allowing Palestinians who want to leave the freedom to do so *not* genocide, but it may be the prevention of genocide.

First understanding

Humanity misunderstands the true causes of our wars, all wars, and that's why armed conflicts occur despite all of man's knowledge and wisdom. Wars are not really about land or resources or even religion. Rather, I believe there is a secret key for unlocking all war amongst mankind, which I call PeaceMatrix™ Entitativity theory.

People don't usually act for the reasons they think they do, especially not as we approach conflict. People usually act based on subconscious impulse, primal instinct, and emotion, justified by select supportive logic after the fact.

Soldiers usually don't start wars, but neither do leaders. The causes for wars that leaders and experts explain to their populations before, and historians explain in hindsight, are materially incorrect. The progressive actions and escalations leading to war are not the *cause* of war, but a symptom of another problem. But if these are symptoms, what is the true problem that is the cause of war?

The key to all human conflict, perhaps more valuable in predicting and preventing war than knowledge of every fact of human history is, that wars are caused by the collective psychological dynamics of in-group versus out-group polarization. As humans approach conflict, an unconscious ideological collective psyche ("hive mind") dominates what we perceive to as individuals to be our rational thought, causing irrationality, strategic errors, escalations, and war.

This is why NATO won't make a peace deal with Russia over Ukraine - because it is not an individual thinking clearly, but a groupthink erroneous collective decision. The hated of Jews is not rational. Nor is Harvard support of Islamic extremists over the "start-up nation" of Nobel Prize winners. Nor feminists supporting Hamas terrorists who would give them no rights. Nor Hamas terrorists aligning with communists who ban religion, etc.

Summary: The collective hive minds that control mankind as we approach conflict do not follow their proclaimed ideals or values, but operate in the collective unconscious under a different set of rules. The word "Entitativity" literally means the degree to which the individual versus the collective hive mind is the sovereign entity.

Second understanding

If we correctly understand these ideological dynamics and find ways to moderate and guide them, we can prevent, avoid, and end wars, maybe all war amongst mankind. By war, I don't just mean nation to nation armed conflict, but civil wars, political upheavals, ideological, group, cultural, and religious disputes.

Third Understanding

The key to the whole mystery of the rules of ideological group dynamics in conflict, and to stopping war is as follows:

Instead of group ideologies being controlled by their proclaimed values and leaders, ideological hive minds grow and spread like separate living organisms, which I call "living organism theory". These ideological living entities behave like amorphous clouds that are the puppet masters of humanity, with power and growth interests separate and superior to their adherents'. They feed off of conflict by gaining more subconscious power over the rationality of their adherents.

This is why ideologies of opposing value systems combine for power. This is why ideological movements historically become more aggressive power movements as they reach their claimed goals - because the ideology becomes more powerful, and even more hungry for more power. This is why democracies are morally aligned with other democracies; not to enjoy being a thorn in the side of dictatorships, but because dictatorships are more prone to conflict as the internal ideology must polarize against an external out-group, leading to expansions and conflicts with other nations. This is why empires fall with often nothing replacing them, because new unconstructive ideologies unite to tear them down for the irrational power gain of doing so. This is why nations believe they are building up their defenses until they feel compelled to attack others. This is also why anti-Semitism grew in response to October 7th before Israel even responded, because the ideological entity was empowered by the primitive brutality and polarization.

The problem:

Our mistake in our efforts to examine, interpret, predict, and stop war is that we think the parties actually want what they claim to want, for example: the Palestinians wanting a Palestinian state. Much of the world treats the Palestinians as though if they only had a state, they would be happy, and there would be peace. But the Palestinians have rejected a state five times, when not a stepping stone to better attack Israel from. They put their "cause" ahead of their own interests.

Our mistake is that we think the individuals are the party to the dispute, or that their leaders are. We neglect to deal with the true negotiator at the table, the ideology, as a separate party to the dispute. Without this factor, our formula of understanding is completely wrong. As we have seen, in 75 years, we've gone from 50% of Palestinians wanting Israel gone, to now near 80%. In Arab societies, ideological influence is regarded as much more powerful than we are used to thinking of them in Western societies.

The correct perspective is that the "Palestinian cause" is a living, growing separate entity. That entity sends human adherent pawns to their deaths by suicide bombing, proving the ideology is the sovereign entity. It survives by continuing the conflict that fuels the living ideological hive mind regardless the interests of individual Palestinians. Palestinians always reject statehood, because if they got a state, the Palestinian "cause" would die, and *it* doesn't want to die. Arafat was threatened with death if he accepted a peace deal at Camp David not by individual adherents, but by the ideology compelling them.

While the world treats Palestinians like a unique culture and people deserving statehood, they are the victims of their own ideological entity and the world's misunderstanding of that. Relocation is not genocide, if the Palestinians choose it, and also because they are arguably not a unique nation, race, ethnic, or religious group, but an arm of a broader Arab ideology, which is why they cannot deal in their own behalf in the first place. They speak the same language, have the same religion, and pray daily toward the same Capitol in Mecca as the people in 22 other Arab states with nearly identical flags.

If hypothetically, Palestinians did get a state, the living ideological entity would become more powerful and even more hungry for more expansion, with an even greater need to expand and attack Israel. Even if Palestinian leaders wanted peace, ideological dynamics would replace them by channeling political and influential power to more extreme leaders who will increase arms and population buildup to attack Israel with. Eventually, the compulsion to attack Israel would be irresistible, and there would be more war, not less. The Palestinian "cause" has already changed from a goal of two-state statehood, to "from the river to the sea" right before the world's eyes.

And hypothetically, even if the Palestinian entity conquered Israel, it would simply have an even greater need to expand and threaten Jordan, Egypt, Lebanon, and eventually Saudi Arabia. The Palestinian "cause" would simply change again right before our eyes from dominance of the Holy Land to regional dominance and a caliphate.

A solution:

The way to deal with the problem and help Palestinians is not to empower an aggressive, and oppositional ideological entity of which its individuals and leaders are merely pawns. Rather, by the type of Entitativity modeling I am developing, to moderate the ideological entity's limitless hunger for expansion and conflict, by:

a) a balance of competing unarmed political parties within a structured system to limit the growth of the "Palestinian cause", with a threat of relocation so the ideology bears a consequence for not accepting the system. Then, the entity is not empowered by external expansion and conquest ideology. Leaders will have to compete to improve Palestinians' lives instead of just declaring war, or else Palestinians will voluntarily leave, weakening those leaders. In essence, you use the threat of relocation to give a peaceful two-state solution the best chance of success.

Or otherwise;

b) if the entity does not accept moderation in its current location, the consequence is relocation to a Palestinian state in the Sinai, Jordan, Qatar, or other Arab land where it is not a stepping stone to try to destroy Israel. Having another location for a Palestinian state would separate the ideology that wants statehood from the ideology that wants to destroy Israel, and trigger debate between them like letting steam out of a steam engine. Thereby, the world is no longer united by anti-Semitism feeding support to the entity. (For example, Egypt and Jordan can treat Palestinians in ways Israel cannot get away with without international outrage and legal and political repercussions.)

Without an option for relocation, Palestinians and Israelis are two ideologies forced to be together like two roosters put into a small cage that have no choice but to fight. But with such relocation options, the "cause" would no longer be a united Arab versus Israel cause, but an internal Arab matter. The result of making it an internal matter is that ideological polarization would be reduced, weakening the entity, making it more manageable, thereby reducing conflict, and helping Palestinians.

If the Palestinians do not accept a Palestinian state elsewhere either, the consequence then is the entity will be weakened further, as Palestinians can be absorbed into other states' populations as refugees. If countries like Egypt or Jordan, or even Iran and Turkey, are truly intent on helping the Palestinians, they can take in Palestinian refugees. The world has no right to force Palestinians to stay in a conflict, nor place the burden and dispute on Israel, when the issue is far less polarized if fellow Muslim states deal with it.

By creating a structured path of cascading weakening of the entity, you moderate the ideology, instead of trying to deal with leaders who are themselves victims of their more powerful ideology. It's a win-win because either you weaken the aggressive ideological grip on the population, or obtain the next best alternative for the people. All "international law" rulings to the contrary are wrong, and this is right, because what is truly immoral is a global mindset and imposed policy preventing Palestinians who want to relocate from doing so.

When our conflict resolution efforts adopt these understandings, ideologies in conflict will stop trying to primitively destroy each other triggering their defense mechanisms and greater polarization and wars. Humanity can stop fighting our conflicts short-term and piecemeal, and address broader and long-term issues coming 10, 20, and 50 years ahead.

We will be able to allow all beautiful cultures to preserve themselves if they choose. Ultimately, humanity can learn to carefully and gently guide an ideological entity in a manner that strategically weakens it for not moderating itself, so it may do so. And, so it may not just grow, but develop, just like we would want for any other living organism.

Egypt Must Accept Refugees at Rafah Per International Law

Israel is not keeping Palestinians in an "open-air prison", the whole world is, with a nonsensically global embargo on Palestinian refugees who otherwise may voluntarily wish to leave Gaza and seek asylum elsewhere. This Arab-pushed policy of not accepting Palestinian refugees, contrary to world's policy of helping all other refugees relocate, oppresses the Palestinians by forcing them to stay in a conflict many may not wish to be part of. Without the freedom of Palestinians to leave Gaza, their leaders have no accountability, and civilians remain helpless pawns of external forces. But world's dark, inexplicable, and backwards policy of forcibly keeping Palestinians in the conflict zone between Israel and Hamas does have a crack of sunlight: The Rafah border with Egypt.

And while the Arab narrative is that Palestinians don't want to leave Gaza, if that were true, no global embargo on fleeing Palestinian refugees would be necessary. And, the long lines at passport and border agencies in Egypt would not be around the corner with applications by family members on behalf of Palestinians begging to be set free into Egypt.

Under International law, Egypt cannot turn away Palestinians who seek asylum and entry at its Rafah border with Gaza as refugees, temporarily or otherwise. Egypt's refusal to do so, and instead, its building up of border fences to keep Palestinian refugees from seeking asylum, is illegal. Allowing Palestinians who wish to leave to do so can ease pressure off the conflict helping all parties involved.

Egypt is a signatory to the United Nations Convention Relating to the Status of Refugees, adopted in 1951, with accession on May 22, 1981. Egypt is also bound by the Protocol Relating to the the Status of Refugees adopted in 1967, and numerous resolutions on refugees and asylum-seekers apply to Egypt, including General Assembly Resolution 2198 (XXI), and the UN Declaration on Territorial Asylum in 1967 (GA Res.1400 (XIV), 21 September 1959).

Palestinians are "refugees", under Article 1A of the 1951 Convention Relating to the Status of Refugees, which defines a refugee as someone who, *"is outside his or her country of nationality or habitual residence; has a well-founded fear of persecution because of his/her race, religion, nationality, membership in a particular social group or political opinion; and is unable or unwilling to avail himself/herself of the protection of that country, or to return there, for fear of persecution."* Palestinians in Gaza have been dislocated, they are not in their country as there is no Palestinian state, they fear persecution, whether from Hamas or the consequences of the conflict, and are unable to avail themselves of the protections of Israel due to Israel's limited resources and its ongoing conflict with Hamas. Regrettably to Israel, innocent Palestinians are caught in the middle of the conflict. And while some leaders in Israel may wish to relocate Palestinians for their own safety and well-being during this conflict, many in the world insist on keeping the Palestinians in harms way, and instead call for a ceasefire that would help Hamas re-arm and re-group to be able to cause more conflict.

The United Nations Relief and Works Agency for Palestine Refugees in the Near East (UNRWA) has also admitted that the Palestinians are indeed refugees. It has defined refugees to include *"The descendants of Palestine refugee males, including adopted children"*, totaling *"5.9 million Palestine refugees"*, of which UNRWA says 1.7 million are in Gaza.

The 1948 Universal Declaration of Human Rights provides that *"everyone has the right to seek and enjoy in other countries asylum from persecution"* (article 14), and as such, the Palestinians have that clear right, to seek asylum in Egypt. (Universal Declaration of Human Rights, GA Res. 217 A (III), 10 December 1948.)

Egypt may not discriminate against refugees because they are Palestinian or from Israel per Article 3 of the 1951 Convention, entitled "nondiscrimination", which states, *"The contracting states shall apply the provisions of this Convention to refugees without discrimination as to race, religion, or country of origin."*

At the Rafah Border, Egypt is bound by the principle of *"non-refoulement"* as defined in Article 33 of the 1951 Convention. Article 33 of the Convention, entitled *"Prohibition of expulsion or return refoulement"* states, *"No Contracting State shall expel or return ("refouler") a refugee in any manner whatsoever to the frontiers of territories where his life or freedom would be threatened on account of his race, religion, nationality, membership of a particular social group or political opinion."* This includes Egypt being prohibited by binding treaty from turning away Palestinian refugees seeking to leave Gaza and enter Egypt at Rafah fleeing danger from Hamas or the ongoing conflict. Whether a Palestinian's political opinion is pro-Israel, or pro-Hamas, they are in danger from the circumstances of the conflict, even if Israel is trying its best to limit that danger to innocent civilians.

The French word *"refoulement"* refers to *"faire reculer"*, which in refugee law *"à la fois l'éloignement du territoire et la non-admission à l'entrée."* (See D. Alland and C. Teitgen-Colly, *Traité du droit d'asile* (Paris: Presses Universitaires de France, 2002), 229) This translated, is *"both removal from the territory and non-admission to entry"*. Thus, Egypt under Article 33 is prohibited from denying admission to entry to Palestinians who show up at the Rafah border seeking asylum.

Because the Palestinian refugees are not being accepted anywhere else, Egypt turning away those seeking entry would constitute sending them back into the conflict zone with no other means of escape. As such, here, rejection at the Rafah border would automatically lead to return to the country where they fear persecution.

The Vienna Convention on the Law of Treaties says treaties must be interpreted in good faith. (Article 26 of the 1969 Vienna Convention on the Law of Treaties) As such, Egypt must allow entry and give asylum until it is determined whether Palestinians seeking asylum are deserving of asylum. Otherwise, Egypt would be denying the ability to determine their obligations under international law.

Additionally, UNHCR has declared that "*A State presented with an asylum request, at its borders or on its territory, has and retains the immediate refugee protection responsibilities relating to admission, at least on a temporary basis. This responsibility extends to the provision of basic reception conditions and includes access to fair and efficient asylum procedures.*" Convention Plus Issues Paper submitted by UNHCR on addressing irregular secondary movements of refugees and asylum-seekers, FORUM/CG/SM/03, 11 March 2004.

Regardless what Israel is doing, Egypt has an obligation to allow entry to those Palestinians fleeing the conflict in Gaza and requesting asylum and entry into Egypt. Egypt may not even delay their entry to try to deter refugees, and must also allow them to stay once inside Egypt without penalty. Per Article 31, "*refugees unlawfully in the country of refugee*", "*The Contracting States shall not impose penalties, on account of their illegal entry or presence, on refugees who, coming directly from a territory where their life or freedom was threatened in the sense of article 1, enter or are present in their territory without authorization, provided they present themselves without delay to the authorities and show good cause for their illegal entry or presence.*" Thus, even if Palestinians enter illegally into Egypt, Egypt's obligations still stand under international law.

This has nothing to do with whether Israel is expelling any Palestinians or not. It is solely about the rights and status of those Palestinians who voluntarily wish to leave Gaza, temporarily or permanently, and Egypt's obligations to accept them. Egypt's obligations are independent of any other aspect of the conflict or peace process. Freedom of movement for Palestinians may lessen the conflict and help the peace process, by not furthering an ideological "cause" at the expense of the individuals who wish to leave and no longer be human shields for Hamas. It is a common sense step, as any solution necessarily requires putting the interests of Palestinian individuals ahead of external agendas.

Egypt also cannot blame Israel for Palestinian refugees wanting to leave Gaza, nor for its own obligations under international law. Arab states have refused asylum to Palestinians to continue the conflict against Israel with full knowledge that Hamas' cause seeks genocide against Israel. Egypt is well aware Hamas is smuggling weapons into Gaza seeking Israel's destruction. And, neither Egypt nor other Arab states have fulfilled their responsibilities of educating the Palestinians for peace, or promoting acceptance of Israel. Arab states have failed to act in the best interests of the Palestinian people, but instead furthered various versions of the Palestinian "cause" at the expense of the Palestinian people. Egypt is not only violating the rights of Palestinian asylum-seekers now by refusing them, but has been for years the conflict could have been lessened. Egypt should demand that the world assist it with resources, and that Jordan, Lebanon, Syria, Qatar, and Saudi Arabia also take in their fair share of Palestinian refugees who wish to leave Gaza, or else Egypt may have to take them all. Yet another reason Egypt should try even harder to get more hostages returned to end the conflict sooner.

It's time we as a world, desiring to help, apply the same standard to Palestinian refugees that we do to all other refugees. Help them. Help the individuals seeking asylum. And, not force upon the Palestinians a genocidal "cause" at the expense of the men, women, and children who wish to leave the conflict zone. Maybe the first step to finally getting Palestinians to act in their own best interests, is for the world to do so.

Recognize a Palestinian State? How about in the Sinai

As the US threatens recognition of a Palestinian state to deter a Rafah invasion, a Palestinian state in the Sinai, even if fully opposed by Egypt, is such a powerful idea as to be potentially unstoppable.

Israel's future is on the line, and as more of our brave young people die in the conflict, we owe them a strategy smart enough to justify their sacrifice. By understanding and addressing the ideological roots of the conflict, Israel might just reshape reality.

The Biden administration announced in January that the US State Department is reviewing options for unilateral recognition of a Palestinian state, as part of a deal for normalization with Saudi Arabia. Blinken was in Egypt February 6 meeting with Egyptian President El-Sisi, and King Abdullah of Jordan visited the White House to discuss the matter on February 12. But Israel's goal, unity to address the threat from Iran, may not be the Administration's. The US inquiry is perplexing, as clear US policy for decades has been to oppose U.N. recognition of a Palestinian state and even oppose all bilateral recognition, absent mutual agreement with Israel. It is also odd, as the Saudis have recently said a US commitment to the political process would suffice. If you'd like to take a guess where such radical US policy ideas are coming from, American exploration of unilateral recognition last occurred during fiercely anti-Israel Obama Administration. And interestingly, Obama was also present at the White House for King Abdullah's visit, for those connecting the dots.

PM Netanyahu's response was rightly, "Israel will not submit to international dictates", explaining such recognition would be a reward for October 7th, and cause more conflict. But what should PM Netanyahu's chess move be if the US regardless announces unilateral recognition of a Palestinian state, or uses such to pressure Israel on Rafah? What leverage does Netanyahu have to try to deter such a unilateral move in the first place? And what solutions exist to this continuing conflict?

Let's put aside that it is nonsensical to unilaterally recognize the fictional impossibility of a state that can't agree on its own borders, is run by an outside-funded terrorist group, and based on external ideology that wants genocidal war against Israel. The US unilaterally recognizing a Palestinian state on Israeli territory would be as nonsensical as Ukraine recognizing a Ukrainian state up to Moscow. Recognizing a state on territory it doesn't control or with borders even Palestinians don't agree on would only trigger more conflict, if not restart the 1948 war all over again. According the the Treaty of Westphalia (1648) which paved way for the modern international nation-state system, the objective minimum requirements for Palestinians to be a nation are absent, including one governing authority's control over the population, and clearly-defined borders.

Let's also put aside that such threatened US policy lacks understanding that any Palestinian state in Judea, Samaria, Gaza, and East Jerusalem, would only be used as a foothold to acquire arms and territory to better attack Israel.

Let's step back and look at the fundamentals of why.

Can Israel destroy Hamas?

Israel has been fighting Hamas for over 35 years, and Hezbollah and Islamic Jihad even longer. Likewise, the US fought the Taliban in Afghanistan for 20 years. We've learned nothing, and the world condemns Israel now more than ever.

Israel is NOT fighting an individual (like Hitler).

Israel is NOT fighting a religion.

Israel is NOT fighting a nation.

Israel is fighting an ideology.

Regardless that Israel is justified, it can't destroy Hamas with force alone because it is an idea. When Israel kills terrorists, new ones take their place. Even if Israel killed every Hamas member in Gaza, new members will sprout in the fertile soil of the Palestinian collective mindset. With Amalek, the Torah told us to "destroy all *memory* of them", but the Palestinians and their leaders in Qatar and their supporters in Iran will remember Hamas. If not following the Torah, this strategy alone won't work.

Israel is actually fighting an ideological entity called the "Palestinian cause." Individuals players, and even leaders, are not independent-thinking actors, but adherents carrying out the will of the ideology.

The correct course is to understand that the only way to destroy an idea is with a better idea.

A two-state "solution" with a Palestinian state in Judea, Samaria, Gaza and East Jerusalem cannot be successful because it would only be seen by Palestinian ideology as a stepping stone toward destroying Israel, to bring in millions of people, better arm themselves, and eventually attack. There are no Arab democracies in the Middle East, and nobody can explain why a Palestinian state would be the first. And with no internal political process, all political gain would derive from demonizing Israel, continuing to advance the same "Palestinian cause", and more conflict against Israel from the better attacking position of statehood. Even if you demilitarized the Palestinian state, Palestinians would simply continue to seek to destroy Israel through better smuggling weapons, population expansion, legal, political, diplomatic, and other means. Empowering them with statehood without first addressing their ideology and its external support like from Iran will only make the ideology more aggressive.

But what if the current US administration does it regardless?

Netanyahu might deter such a move, or respond, with something equally unconventional, and much smarter:

A key to the whole ideological problem in a nutshell:

In sum, the "Palestinian cause" is an ideology. And, the only thing that can defeat an idea is a better idea.

How?

The Palestinian "cause" ideology is actually made up of two separate ideologies:

A. one ideology that simply, innocently, wants to help Palestinians live better lives in peace, and,

B. one ideology that wants to destroy Israel.

The problem is, they are merged. As Israel recognizes, when you feed one, you feed the other. The world thinks its helping A, but it's also helping B. Globally, those who want to simply help the Palestinians are united with those who want a Palestinian state as a stepping stone to destroying Israel. Because this is on a powerful, emotional, ideological level, Israel's logical explanation is ineffective at persuading most other nations. These ideologies unconsciously control the purportedly-well-reasoned viewpoints of individuals and leaders and journalists worldwide. These ideologies span across the globe, inhabit the views of many groups and leaders of nations, and run down the middle of the mindset of each individual. Most are unaware whether deep down inside their subconscious they support A or B, or both, or 80/20, or 90/10, etc. After all, at what point in the human unconscious is supporting A and simply not being too concerned about B actually supporting B also?

Another way to see it is: extremists can only swim in a sea of moderates. This is why recognizing a Palestinian state without defining borders, or without their recognition of Israel, is the exact opposite of what will create peace. It will unite and empower the two parts of the ideology even more. When ideologies become more powerful, they don't tend to become more peaceful, they become more aggressive.

The key to any solution is:

Divide the ideology. Take actions that divide the Palestinian "cause" into its sub-ideologies, separating them. You can't fight an ideology just by killing its members, but if smart, you can get an ideology to divide and fight itself, whether through debate, policy, politics, economic channels, other otherwise. Israel has done this in part by turning the Palestinians against Hamas to some degree, but this is insufficient long term.

The solution is to propose policies and take actions that help A **at the expense of** B. And then, create the systems that irresistibly empower and enrich those who support A at the expense of the interests of those who support B.

Turn: A + B = conflict with Israel

Into: A gains as B loses

Then: A + Israel = isolates B

Here is one outside-the-box example that admittedly sounds crazy:

Recognizing a Palestinian state in the Sinai

Of course, everyone will instantly reject such an unconventional idea that goes against every Arab country's longstanding policy. But, this strategy might work even despite everyone rejecting it, as a good idea based on truth can take hold even if most of the world fights to stop it.

Israel declaring recognition of a Palestinian state in the Sinai is no less crazy than the US declaring a Palestinian state in Judea with a capitol in East Jerusalem, or any nation declaring part of any other nation or their capitol as belonging to a third party. If it sounds crazy, Egypt has not only reinforced its barrier at the Rafah border, but is already creating a fenced-in area across the border in the Sinai for expected Palestinian refugees.

Netanyahu's chess move, the US and Egypt should know, in response to unilateral recognition of a Palestinian state in Judea and Samaria, may be Israel's recognition of a Palestinian state in the Sinai. It may also be used as a threat to Hamas, that if they don't release the hostages immediately, Israel may recognize a Palestinian state in the Sinai.

How in the world?

A Palestinian state in the Sinai separates the two sub-ideologies that make up the "Palestinian cause". It is more outside-the-box than anything tried in the last 75 years, because it divides and puts the two ideologies against each other. It creates a new alternative, fractions, and a situation where one ideology gains as the other loses.

With a Palestinian state in the Sinai, Palestinians can have their state that they and the world are demanding. They can have borders with Israel and Egypt, a seaport, an airport, free trade, an economy, a fertile Mediterranean coastline, waterways, and global financial investment and support. It would be near the major city of Cairo, it would be near Aqaba Jordan, and near The Line - the new Saudi Arabian modern city being built. The Sinai is 60,000 square kilometers and most of that land is unused, whereas Gaza is only 365 square kilometers, meaning that a small fraction of the Sinai would give Palestinians more land than Gaza, Judea, and Samaria combined.

The only thing missing is, it doesn't bring the Palestinians closer to destroying Israel per the goals of Hamas. That's a heck of an argument to try and rationally win when the whole world has been pounding their fists for a Palestinian state and an end to their suffering, all of which they would get. It changes the dynamic away from a zero sum game with Israel. It furthers the "Palestinian cause" of improving the lives of Palestinian people with statehood and autonomy, while weakening the "Palestinian cause" of destroying Israel. The Palestinians who want a state and peace will fight for it, and those who only want to destroy Israel will fight against it. Thus, they will debate and fight each other. That debate will go global, and Israel would recruit millions of people and entire nations who ask "why not" to its side. One Palestinian cause would strengthen, and the other weaken. For every argument that a Palestinian is suffering, they now have another option to move to a Palestinian state in the Sinai, ending the ideological stalemate that blames only Israel for whatever problems Hamas can cause. The issue is not whether Palestinians or Egypt reject it, but transferring the blame against Israel if they do. Israel can offer a state and a solution, without rewarding Hamas at all.

It's not Israel kicking the Palestinians out, but the Palestinians choosing to build a state where they can do so while weakening their terror breeding ground so they can govern themselves. And if you don't think that idea is powerful, wait till you see the viral videos of Palestinians at the Rafah border demanding food, water, and medical care for their children being turned away by Egyptian authorities.

Criticisms:

- It's a legal impossibility. - True, but so is external recognition of a Palestinian state on Israel's territory, or without clear borders or leadership.

- Egypt and Arab states won't agree. - True, but they don't have to agree. As Hamas continues the conflict hurting the Palestinian people, many Palestinians will flood into the Sinai to receive international aid and better life than in Gaza, boosting the ideology.

- Egypt has threatened withdrawal of its Camp David Treaty with Israel. - Egypt has no basis to complain about its own duties under international law to accept refugees at its border who seek asylum. As detailed in my other article, Egypt has a duty to accept refugees per its legal obligations as a signatory to the United Nations Convention Relating to the Status of Refugees, adopted in 1951. In fact, Egypt's failure to take in refugees for decades to advance the "Palestinian cause" ideology, and allowance of smuggling into Gaza, is arguably partly to blame for the conflict. Once Egypt has enough Palestinian refugees, they will want the same thing there, their own state. Egypt will seek international involvement on the issue, and to distribute its responsibility to other nations. With Egypt facing severe financial problems recently, other nations will be able to push Egypt to adopt measures in exchange for funds. Ultimately, Egypt could benefit with a new additional neighboring state as a trading partner and economic stimulus in the Sinai, and the entire Sunni Arab world can benefit from uniting with Israel against the threat posed by Iran. While this strategy doesn't address the Iran problem, it empowers Israel with new ideas to unite with Arab allies to deal with it.

- The international community will be angered at Israel. - True, but they will also have a new way of helping Palestinians without hurting Israel, lifting pressure currently united against Israel. The international community can pump billions of dollars into the new state in the Sinai, including frozen terrorist and Iranian funds. Besides, if Palestinian people want to go elsewhere, who are foreigners to dictate that they can't?

- Hamas will take over and or sabotage the new state. - This would happen even more likely with a state alongside Israel. This way, at least the international community could get involved, help moderate Palestinians, facilitate a political process, and Israel could support it as it wouldn't all be at Israel's expense. It would align the efforts of Israel and much of the world.

- Such a state would collapse. - That could happen alongside Israel anyway. Then, such populations of refugees could be absorbed into other Arab states, which both Israel and Egypt would then be united proponents of.
- Such a state would still be a threat to Israel. Ideologically it would be less so, and it would have the freedom to prosper, thereby at least able to build a constructive ideology.

Whether such an idea would ultimately work is up to the Palestinians and Arab world. If it comes along with an ideology of respecting Israeli sovereignty in exchange for Arab prosperity, it can work. If it doesn't, it'll still reduce the mounting global pressure on Israel.

Such a policy would also follow the scriptures of all three major Abrahamic religions which provide for Israel's right to the Holy Land. This includes the Qur'an's Zionist language giving Israel to the Jewish people: "Remember when Moses said to his people: 'My people, remember Allah's favour upon you when He raised Prophets amongst you and appointed you rulers, and granted to you what He had not granted to anyone else in the world." (Qur'an 5:20). "My people! Enter the holy land which Allah has ordained for you; and do not turn back for then you will turn about losers." (Qur'an 5:21) "And thereafter We [Allah] said to the Children of Israel: 'Dwell securely in the Promised Land" (Qur'an 17:104). Every other Arab state that doesn't try to conquer Israel prospers consistent with this language, and it's about time the Palestinians did likewise. Developing Arab communities around, but not in Israel, has basis in scripture to create ideological support for a solution that fully accepts Israel. This policy provides financial and long-term ideological support for that solution.

In sum, Israel's challenges are not conflicts against an individual, nor against a group. They are against an ideology. That ideology tells Arab leaders what to think and do, and it can be divided. When we start strategically thinking about how to divert ideological momentum instead of killing its adherents, that's when we start diverting the river instead of using a teaspoon, and taking our first steps toward peace.

UAE's Egypt Land Deal Avoids Peace at Any Cost

Egypt's recent financial troubles with its national debt approaching its GDP created the perfect opportunity for resolution of the entire Palestinian conflict. Nations could have pressured Egyptian President El-Sisi to allow Palestinian refugees through the Rafah border per Egypt's obligations under the United Nations Convention Relating to the Status of Refugees (1951) per my other articles, to cooperate for release of Israeli hostages, and even build a beautiful, large, prosperous Palestinian state in the Sinai. Everyone would be happy.

But no, we are not that smart.

Of all the missed opportunities in the current Gaza conflict, few are as expensive as the UAE's and Abu-Dabi-based ADQ investment company's $35 billion dollar land investment in Egypt's failing economy. Instead of the wise nations of the world cooperating to use that $35 billion to invest in the Sinai and develop Egypt's mediterranean coast there sufficient to house an entire Palestinian state and resolve the entire conflict, the funds appear intended to accomplish the precise opposite. They appear to be intended to bail Egypt out of its financial problems, by developing Egypt's other, western mediterranean coast, its Ras El-Hekma project west of Alexandria. Of course, that amount invested in a Palestinian state there would profit many times the return. But no. The funds appear to be a bailout of Egypt, intentionally so Egypt is not pressured to help Israel do anything that might alleviate the conflict. BP's $1.5 Billion dollar investment in Egypt will help along similar lines after Egypt's recently lower oil production.

If the estimate of $18 billion total international aid given to the Palestinians since 1948 is correct, this is just about double that amount. While the investment is enough to build an entire Palestinian state on the Mediterranean coast designed with modern, world-class homes, resorts, and infrastructure, instead the investment is apparently designed instead to avoid peace at any cost. As close to a peaceful resolution as the world could get while still mocking it, the missed opportunity is like running a

26-mile marathon only to run head-first on purpose into the pole at the finish line.

If aliens exist, they are looking down at us shaking their heads.

Adventures of a Young Mashiach

How Mashiach might see the collapse of the United Nations

Here is one interesting interpretation of what Mashiach's adventure may first look like when he comes, so we may as well start asking these questions while we wait for him. With divine hope and creativity is how we should be looking at all our problems anyway.

In this three-part piece, several mysteries regarding Mashiach may be answered, including:

- How might there be world peace one day?
- How and why would there be a new world government in Jerusalem?
- How could we ever see the decline of anti-Semitism?
- How could everyone on earth have a divine perspective?

I definitely don't claim to be Mashiach, for I would be "riding a donkey" indeed (Zechariah 9:9). I am a shy, quiet, unimportant, introvert, who probably couldn't get a dog to listen to me. I can just imagine old Jewish ladies sarcastically saying, "Ah, that's all we need." In a book I wrote, I joke that if I ever do accomplish anything important and have a party to celebrate, people would gently push me aside as they make their way through the crowd looking for the most important person in the room. I didn't grow up observant and didn't even start getting close to Judaism until after age 40. And, I have zero interest in power or honor or attention whatsoever. I just want peace and quiet. But I thought this was a cute title and playful writing trope considering the fascinatingly-converging topics. So, I write this satisfying my soul yet taking comfort in the consolation that probably nobody will read it.

There are some good ideas in the Bible about Mashiach, like his bringing world peace, "Nations shall not lift the sword against nation; neither shall they learn war anymore." (Isaiah 2:3, Micah 4:3). Every person will have a divine perspective, "The earth will be full of the knowledge of God, as the waters cover the sea" (Isaiah 11:9). There will be a world government in Jerusalem, "Jerusalem will be called the City of Truth" (Zechariah 8:3). And, the smallest tribe will be mighty. (Isaiah 60:22)

With the U.N. condemning Israel more than any other nation, the Human Rights Council alone doing so 45 times since 2013, more than all other nations put together, and conflicts not surprisingly escalating nonetheless, Mashiach would be sure to face many challenges on his journey in making world peace come true.

Our creative sci-fi imaginations have apparently thought of every realm of science fiction conceivable; from space aliens, to traveling to other planets and galaxies, time travel, other dimensions, futuristic technologies and worlds. But in all our imaginations, there hasn't been a hint, of an inkling, of the idea, that someone may invent a radically new way for humanity to better make peace amongst ourselves.

Well maybe now we have.

If this is the first you're hearing of the conceptually making peace amongst all of mankind, buckle your seat belt.

I don't know when Mashiach may come, but I can tell you a little about what the start of his adventure may look like, as I've also been thinking about world peace. Writing about Israel, political, and global security matters, and having difficulty organizing my own thoughts, I began seeking ways to organize the elements of the political and ideological disputes. As I continued to meditate on these concepts, they began to increasingly look like a new type of peace-building system. Having a child-like curiosity, my options are to assume they are worthless, or explore their usefulness and pass on what I learn. I could shout my ideas from rooftops hoping dogs bark back, or I write about them. So here is some of what I have discovered.

As background, I have written about several important related concepts. I wrote "Humanity is in its Infancy in Understanding War", explaining that humanity is primitive in our understanding of, and our systems designed to prevent war. Wars don't happen for the claimed excuses. In "The End of Moral Equivalence with Israel", I explain why the world's view of the current conflict is logically flawed. And in "The Trail of Mistakes to Nuclear War", I explain more broadly why parties in these conflicts are making mistakes. Finally, in "The Metaphysics of Anti-Semitism", I explain anti-Semitism, and its relation to current and future conflicts, as well as hidden clues about human nature that lie in the unconscious hatred of the Jewish people, that if better understood, might help solve all hatred amongst mankind.

Wars occur because the world develops unanswered questions that result in in-group versus out-group polarization, along unconstructive poles no less, and ideologies take on a life of their own. Nations end up going to war often without even knowing the correct question at issue.

But imagine we could dissect the unconscious reasons why humanity goes to war with good ideas and questions. An idea can be the most powerful thing in the world, but even before that, a question to find such an idea can already start to reshape the world. It is such questions that a young Mashiach will one day ask, and we should be asking until he gets here, because otherwise, we are headed for wars:

The bad news:

Historically, our top-down global systems of leadership of nations by individuals resulted in good kings and bad kings, good times and bad times for human history. But in the age of nuclear weapons, trial and error global governance won't be sufficient. We have ideologies that supersede leaders, sewing the next rounds of future conflicts already in the pipeline.

The world's response to October 7th showed us that the same hatred for Jews I explain in "The Metaphysics of anti-Semitism" that arose against the Jewish people in countless localities for millennia, may one day envelop the world in our current primitivity. The U.N. General Assembly recently voted for Israel to ceasefire in its war to destroy Hamas with 153 nations in favor, just 10 opposed, and 23 abstentions. The woke youth of today ages 18-24 who don't believe Israel should exist according to a poll by Harvard-Harris December 2023, will become the leaders and decision-makers of tomorrow. One day, Israel will be virtually alone and the U.S. will not be giving its near-sole support. Israel will be correct and nearly the whole world wrong, but it won't matter. Anti-Semitism is not logical, but will regardless grow into a world war and nuclear war, if not this time, soon. Most simply explained, as new ideologies grow and morph, they have a proclivity to destroy their foundational morality to continue to survive. And no foundation is more necessary for new divergent ideologies to attack, than Western Civilization's foundational Judeo-Christian values.

The faults with the U.N.:

UN Secretary General António Guterres told the United Nations at the opening of the 77th General Assembly, *"Our world is in peril and paralyzed,"* and *"We cannot go on like this."* He warned, *"Our world is in big trouble"*, that *"Divides are growing deeper…"*, *"challenges are spreading farther,"* and of *"colossal global disfunction."*

As such growing new ideologies distort mankind's "logic", soon the "best" "wisdom" of the combined United Nations will be demanding that the Jewish state, the source of Western Civilization's morality, be weakened so that the last 0.4% of the Middle East can become a 23rd Arab state, because 99.6% of the Middle East is not enough for them. The U.N. can barely pass a resolution against Iran or North Korea, but can easily pass 15 resolutions a year against Israel, often with the U.S. as Israel's sole defender vetoing Security Council Resolutions.

In 2020, the United Nations General Assembly, the forum of every member nation, 193 countries, passed 17 resolutions condemning Israel, the only democracy in its region, and only 6 against the entire rest of the world. The 2021-2022 session will likely have 14 condemning Israel, and 5 the entire rest of the world.

Tragically, the U.N. finally made itself obsolete in 2016, when it declared that the 3000-year-old Jewish City of Jerusalem and Holy Jewish Temple doesn't belong to the Jews, when the U.S. (under Obama secretly pushed for and) abstained, and allowed Security Council Resolution 2334 to pass. Not to mention Judea, which I like to say, sounds awfully Jewish.

The U.N. failed to prevent the Russia-Ukraine war ("operation"), nor is able to prevent a potential China-Taiwan war ("reunification"). Nor can the U.N. stop the Gaza war, nor Iran's nor North Korea's nuclearization, despite being contrary to the Nuclear Non-Proliferation Treaty (NPT). Is the world going to rely on a system that requires wars be "proportional" so they last forever? A system in which the wisdom of most of the people of the world have virtually no say? A system in which the single Islamic nation has up to 57 votes, and the source of Western morality, Israel, has never been allowed to even speak at the U.N. Security Council? Not to mention, the U.N.'s reliance on our nation-state system of international law, which facilitates terrorist groups like Hamas we didn't have pre-U.N.. Terrorists escape accountability like we're a bewildered audience when a magician pulls a rabbit out of a hat, not knowing it's Iran.

The combined output of the world's best minds is only as good as they system they operate in. Genocides are occurring all over the world. Now, Iran's anti-Semitic and genocidal leadership that oppresses its own people is in a breakout for 6-10 bombs with which they admit they plan to "wipe Israel off the map". After that, they will nuclear blackmail the rest of the Middle East and Europe. Yet much of the world is focused on ganging up on Israel militarily, politically, and legally in response to her defensive war in Gaza after the October 7th massacre, while Israel can't even table its own General Assembly resolution in the U.N.. Now, South Africa has accused Israel of genocide in the International Court of Justice (ICJ) in the Hague without looking at all the other ongoing genocides, including against Israel. In reality, many are just ideologically aligned against Israel, which causes them to see the world inaccurately. Hand-selecting only the Jewish state to be put under the microscope while it fights to recover still-kidnapped Jewish children not only goes against natural law. Simply hearing the case of the Jewish state and only the Jewish state on trial will destroy the credibility of the court. The problem with international law is that if the "deciders" become ideological and lose perspective, they destroy the system. Judges can rule that the sky is green and the grass is blue only so long, as a teacher who teaches falsely will eventually lose all his students.

In 75 years since Israel's founding, the world's moral perspective has shifted from supporting the Start-Up Nation to now supporting the terrorist-led Palestinians. Just imagine if a mugger snatched your wallet, and while you were struggling to take it back, the world changed its mind and passed a law making theft of wallets legal. Terrorism went from a scourge on humanity, to having global support. As long as there is political gain from opposing Israel, opposition will grow. And if a young Mashiach were here, he would be asking, "why?" And if you answered that it's anti-Semitism, he would be asking, "why?"

The U.N. and other current peace-building mechanisms are increasingly failing the globe and increasingly turning against Israel. We must improve the current system or invent a new one now, before the next inevitable World War, or cling to this sinking ship of an ineffective system like the League of Nations, which became a debating society that failed to prevent World War II. And by improve, I mean something better for Israel, better for the United States, better for Judeo-Christian values, better for all of Western Civilization, better for every beautiful and unique nation and tribe and culture and religion and people; and better for the whole world.

Of all the reasons Israel is being smacked around like a ping-pong ball in the U.N., the ultimate one is because we haven't thought of something better. But by asking questions, maybe we can indeed find something better. Remember that the only thing that can stop an idea, is a better idea.

Maybe there is an idea that is much, much better.

And maybe that idea is coming.

(To be continued)

The above is, Adventures of a Young Mashiach, Part 1:

The theoretical beginnings of a better global peace-building system is discussed in:

Adventures of a Young Mashiach, Part 2.

Here starts, Adventures of a Young Mashiach, Part 2:

The theoretical beginnings of a better global peace-building system:

The Jewish people, the source of Western Civilization's morality, are increasingly opposed in the world, despite being more moral than most, if not all other nations. The anti-Semitism that arose against Jews in nearly every generation and community we have lived in as a minority, now appears to be going global as our globe becomes a "community." The Holocaust was a killing of Jews across many countries simultaneously cooperating, and on this trajectory, the mental illness of anti-Semitism may soon infect much of the world. The Jewish nation defending its right to exist regardless what the world thinks, as the dear Golda Meir said we should if we must, may not solve a worsening problem forever, without more.

75 years after the United Nations Convention on the Prevention and Punishment of Genocide, of all the nations and their wrongdoing, the Jewish state is on trial for defending herself after the October 7th massacre. And likely, those putting Israel on trial have no consideration for the possibility that their legal actions may instigate more violence. Even though surprise attacked, Israel's favorability in one poll has dropped nearly 20 percentage points since the attack across nearly 50 nations, and Israel now has deeply negative favorability even in Western allies like Japan, South Korea, and the United Kingdom.

If not for the United States' veto defense of Israel over 40 times in the U.N. Security Council in recent years, the global community would have destroyed Israel long ago through a combination of military attack, when that fails: terrorism, when that fails: political attack, and when that fails: boycott, and when that fails: international law. And they will even destroy international law itself to do it. The logical gymnastics of terror victims' families on trial are worthy of gold medals, and yet it makes perfect sense to the accusers. Those who come to see the Jews as evil, themselves become evil, as their moral perspective is skewed by changing ideological phenomena. And they are neither able to learn from history nor look in the mirror to see it.

On our current path, sooner or later, one day Israel may be alone. And even if Israel were no more, the same irrational anti-Semitism would just become irrational anti-Americanism, irrational anti-Judeo-Christianism, and irrational anti-Western Civlizationism. The tide is growing against the people who codified the value of human life. Global opposition is growing against the United States, the greatest contributor to human freedom in history. And soon Western civilization, Mozart and all, may be despised and targeted for elimination to the sound of thunderous applause. Of course, the nukes will be deployed long before than fully materializes.

Our international system is unable to stop the current Russia-Ukraine/NATO war - the largest land war in Europe in 70 years. It is unable to stop the Israel-Gaza-Iran proxy war, the largest Middle East war in a half a century. it is unable to prevent China from invading Taiwan. These moving pieces may be the beginnings of a slippery slope from which humanity may never recover. Eventually, all of humanity may fall into permanent nuclear dark ages, despotism, and savagery devoid of light, unless we fix our current path.

We need something new - something better for Israel and the world.

So until Mashiach comes and accomplishes all that he will, I believe we should try to understand our problems better. Much better.

I woke up from the most amazing dream one beautiful morning, in which the whole world loved Israel, and there was world peace. As I woke up, the feeling of love for Israel and world peace was so thick with realness I could touch it with my hands, and swim through it. There was world peace not despite us, but because of us.

So I asked questions. I asked, "why?" "Why do so many sometimes irrationally oppose the most moral peoples?" "*Why doesn't Israel have peace?*" "Why don't we have world peace?"

And then I asked better questions: *"Why can't the whole planet worked constructively toward solutions together?" "Why can't we as humanity understand ourselves better?" "Why can't the world see Israel's moral perspective?" "Why can't we all see each other's viewpoints and be on the same drawing board?"*

I even asked impossible questions: *"Why solve one major world problem, when we can solve them all?"* And then, *"Assuming there is a way to make world peace, from my limited knowledge and resources, what would that look like?"*

Eventually, silly me, I asked the most impossible question backwards: *"Imagine that all of humanity made world peace, imagine everyone generally got along, and we were able to remain rational and work through our divisions and achieve incredible things, and imagine it was brought about by some kind of human/spiritual ingenuity that accomplished this. What would it look like at its very beginning conception?"*

The PeaceMatrix™ is the answer I got.

Like a curiosity I haven't been able to wrap my mind around, for about four years now I've been wrestling with my beautiful dream and its derivative questions. What if there is a completely different way to resolve human conflicts and make peace?

The question became chains of questions, and increasingly looked like the conceptual framework for a new system of global peace-building with potential to resolve every conflict between every nation, tribe, culture, religion, and political viewpoint, and establish world peace.

What if two miracles that can't be accomplished separately can be accomplished together? Stranger things have happened. Israel playing a role in helping develop and implement a revolutionary new type of global peace-building system to not only solve its own wars but others', could correct the world's perspective to see Israel as a moral center again. Maybe much of the world's opposition to Israel is in the gulf between Judaism being source of Western morality and all the diverging ideologies that seek to create new moral frameworks, and maybe this is the way to fix it.

As we wait for the coming of Mashiach, imagining his adventure one day may be a clue to our path forward now. If you were a hyper-optimist with faith in the ability of God to accomplish anything, you wouldn't just think of solving the Palestinian versus Israel problem, but Palestinians versus the "Palestinian cause" problem, Palestinians versus Arab nations problem, Iranian leadership versus its people problem, Europe versus uncontrolled migration problem, and perhaps every other also, to truly be "a light unto the nations." In my article the Metaphysics of Anti-Semitism, I asked; "What if within the mysteries of why anti-Semitism happens are the secrets to solving all conflicts amongst mankind?" After all, if we could figure out that paradoxical hatred, we could probably figure out them all. Perhaps it's the best we can do until we get to Mashiach, "And he will judge between nations and decide between peoples." (Isaiah 2:2-4, Micah 4:1-3.)

Wanting to solve all wars may seem delusional, but the other option is to proceed towards the world's increasingly nuclearized and uncontrollably-polarized path hoping nobody ever uses them, which is at least as delusional.

I, for one, refuse to accept that the story of humanity is; "fought since the dawn of man, invented nuclear weapons, invented social media, destroyed ourselves, the end."

We must be more rational than that. Perhaps the goal, until Mashiach comes, is to turn the world more rational; less driven by ideological polarization and emotional compulsion; more pre-frontal cortex, less amygdala. Maybe that's what Ezekiel 36:26 means when it says, "I will remove the heart of stone from your flesh and give you a heart of flesh."

It makes some sense that with system that is pro-Israel, pro-U.S., pro-Western, pro-conservative, pro-Judeo-Christian, pro-every unique and beautiful culture, that synchronizes moral perspectives, with Israel, and everyone else, with the jewish people being a key in building world peace, the world might actually fulfill my dream and love us.

By the way, synchronizing moral perspectives is one of the keys to maintaining peace among humanity. If our in-group versus out-group polarization is not determined by (aligned) moral perspective, then it's going to be determined by worse criteria like race or tribal association, short term power gain (think Russia China Iran), or sympathy towards whomever is suffering most in that moment (You guessed it, now Palestinians). Ideally, polarization would be determined by soccer team, or even us all together against a common goal, if I continue dreaming long enough.

As a side note, empathy triggers brain neurohormones like oxytocin, which is activated in the central nervous system with in-group identification and bonding and trust, and which is conversely reduced when dealing with an out-group. With repetition, triggering empathy causes people to be conditioned to side with Palestinians on a neurochemistry level infinitely more powerful than logical analysis of the moral points of the parties in conflict. That's one reason intelligent people side with terrorists. It's also why Jews are only widely favored when suffering. Most countries are troublingly aligned by tribal interest or short-term interest, and increasingly the U.S. alone operates under a moral code as the world's policeman. Most nations are happy to avoid that burden, and just take advantage when convenient.

Without synchronization, eventually the same anti-Semitism-like polarization that targets Jews for our morality will eventually target the United States for its moral positions, and Christianity for its, and eventually all of Western Civilization. And we are already starting to see it.

To make peace, we not only need to win the rational argument, but also turn the world rational enough to accept it.

What is the PeaceMatrix™?

As I struggled to understand and explain human conflicts, I realized that a key reason for war is that we are arguing over the wrong issues. We spend most of our time and energy arguing piecemeal over the inflammatory issues, not the constructive ones. We spend our time trying to convince everyone that the other side is evil, without even scientifically mapping the varying moral perspectives of what "evil" means. Leaders often become bottlenecks warding off good ideas. Our focus is usually during war, not calm when opportunities arise. Points of progress are missed like ships passing because we have no central organization system for the debate, no common ground, no way to advance the discussion constructively beyond the limits of the human character and our current communication systems.

I realized that the way we see wars, through the lens of our media and current events and experts, is insufficient. What we need is precisely a divine perspective.

How did I invent it?

As I struggled to understand and explain the world's conflicts to answer my questions, I kept organizing and reorganizing their elements, again and again. Eventually, it dawned on me that I had categorized them. By properly categorizing the elements of our disputes, we put everyone on the same page, and that is literally more powerful than anyone can imagine. What initially may look like a mind map diagram, or even the best conflict-mapping system on earth, can actually be something far more amazing. Our problem is not lack of information, but inability to properly organize it and give it constructive purpose.

The PeaceMatrix™ is an organization system for the subcomponents of all human conflicts.

The PeaceMatrix™ is a modeling system that creates geometric models of every scenario and viewpoint of every conflict, categorizing and breaking down its elements under 26 primary question categories (A-Z). If parties watch different media, read different books, or have different values or goals, they are nowhere. But imagine a single drawing board for all of mankind to communicate, cooperate, and collaborate in a solution-oriented manner. With all elements of all conflicts in categories (definitions, wants, history, synchronizing moral perspectives), we can channel attention away from inflammatory issues towards the workable details of problems, and their solutions.

Such a system may one day give everyone a divine perspective by letting everyone see all sides of every point of every problem. It resolves cognitive dissonance by presenting all viewpoints; your viewpoint of your side, your viewpoint of my side, my viewpoint of your side, my viewpoint of my side, etc., essentially empowering those with functioning prefrontal cortexes. It can make every tribe mighty if morally correct. And since that morality is objectively based on Jerusalem's 3000+ years of wisdom as a source of Western morality, and invented by a Jew, in a sense, regardless where it is geographically located, the conceptual framework is based on Jerusalem as its moral center. Though, it would be nice if it were in Jerusalem, too.

The closest conceptual work I am aware of to the PeaceMatrix™ is the work of Ken Wilber, who has been called the "*most comprehensive philosophical thinker of our times*" by former New York Times Reporter Tony Schwartz per Ken's book, Eye of the Spirit. Ken Wilber is author of many books including "*A Theory of Everything*", and he is the developer of "*Integral Theory*", a method of bringing together all disciplines and mapping out all human possibilities applicable to nearly any field under a four-quadrant model called AQAL "*All Quadrant All Level*". Those quadrants being self and consciousness, brain and organism, culture and worldview, and social systems and environment. The closest application conceptually to the PeaceMatrix™ that I have found is the application of Ken Wilber's system to conflict resolution in a book called "*Integral Conflict – The New Science of Conflict*", by

Richard McGuigan and Nancy Popp. I was lucky enough to discover both Ken Wilber and his work, as well as Integral Conflict – the New Science of Conflict, only after having already completed the fundamentals of PeaceMatrix™'s conceptual design, framework, and basic implementation strategies. The PeaceMatrix™ is designed to do everything their ideas do, but is superior in many ways as I specifically designed the PeaceMatrix™ for conflict resolution. Most of their developments may largely fit in 1 of 26 categories that make up the PeaceMatrix™ system, the Culture/ideology Chain of the broader, even more mega, meta PeaceMatrix™ system.

What are we organizing?

Since human conflicts are caused by unanswered questions, the PeaceMatrix™ is an organization system of a conflict's outstanding questions. While answers are almost always wrong, good questions open consciousness and the metaphysical universe of infinite possibilities. For example, if someone calls you a bad name, you can either yell back at them and escalate, or ask them their definition and watch their amygdala shut down and their prefrontal cortex kick in as they suddenly look like a deer in headlights starting to think. People who cannot agree on anything else, may still be able to agree on a question to ask. And then, a series of questions. The system asks questions like the Talmud, then asks another question derived of the previous question, and so on infinitely, until we arrive at answers. My parents will be happy to know that my childhood habit of asking questions to the complete surrender and capitulation of all nearby adults may be fully realized.

The PeaceMatrix™ system is designed to create a living, growing puzzle of outstanding questions representing any conflict, combining understandings of psychology, philosophy, history, domestic and international law, geometry, human nature, neurochemistry, and metaphysics. It allows development of chains of derivative questions, meanwhile extracting and hyper-organizing all of humanity's best ideas. We use one puzzle, to solve another puzzle. No more wars. It asks "why" like a child. It's an expansion of global consciousness, for mankind to finally have a divine perspective, to put down the sticks and stones, and finally be able to think. Eventually, every potential and actual conflict on earth may have its own PeaceMatrix™, with the entire world collaborating to solve it. I figured, Jews always love to answer a question with another question, so maybe the way to make the world more peaceful is to make it more Jewish.

The best part is, I don't need anyone's permission to implement it. Not that those running the United Nations nor anyone would ever hand over power to anyone else, let alone me. It's not a power-based system, but a communication system. It's self-implementing. As it grows and helps resolve more conflicts, theoretically, it'll draw more power from ineffective systems, increasingly assisting if not replacing the role of every other system of peace-building. Its use will grow and become unavoidable, because nothing can stop an idea whose time has come.

I know Jews are concerned that any "objective" system will turn against minority Israel like the United Nations, International Courts, international organizations, and so many communities and nations have. Well first, that means we don't have much to lose. But the system I propose is not a majority-vote democracy in which anti-Semitism can spread and dominate. Rather, it shows all viewpoints, including contrary viewpoints, and focus is channeled according to depth of constructive analysis. The world is so obsessed with a misconstrued understanding of "democracy" now, that we forget a majority can be mistaken, by belonging to a single viewpoint that is mistaken. Mark Twain agreed, that "whenever you find yourself on the side of the majority, it's time to pause and reflect."

I know Israelis also have a deep concern of any thing labeled "peace building", because such systems turn Leftist at the expense of Israel's security. My system is designed to objectively re-center the world's moral perspective around the values of Israel, Judaism, the United States, Judeo-Christian values, conservative values of all cultures, and protecting and preserving every beautiful tribe, nation, and people. It has a built-in moral framework that prioritizes quality of viewpoints, rather than political pull or brut numbers, and by a party's contributions to humanity, longevity, indigenousness, uniqueness, history of non-aggression and non-conquest of others, asking good questions, and the ability to accept diversity of perspectives. Whereas a hostile ideology in conflict cannot withstand a diversity of opinion, with one Jew you already have two opinions. So it is objectively a pro-Jewish system.

In short, the PeaceMatrix™ is an infinite global chess game of questions where the opponent is the conflict, not the other party. It's the solution in a world where the logically-flawed ideologies that make war cannot withstand debate, or a multi-perspectived analysis of a few good questions. It's an ongoing perpetual debate in diagram form until the conflict is solved. It's not a power governance system, but a communication system. You don't need an army to defeat another army; you just need to ask them a question they can't answer. As I have said, the opposite of war is nuance.

The PeaceMatrix™ is a metaphysical weapon that destroys the elements of conflict. All human conflict. and I believe it will change the world. Maybe instead of a "peace-building system", I should have called it a "conflict analysis, dissection, chew up, spit out, evaporation, and elimination system."

While the theoretical system is merely in its infancy, I'll show you how it works in the next segment.

End of Part 2

(To be continued in Adventures of a Young Mashiach, Part 3, also known as, "the good stuff")

Adventures of a Young Mashiach, Part 1 & 2 are above.

Adventures of a Young Mashiach, Part 3, starts below:

A demonstration of the theoretical PeaceMatrix™

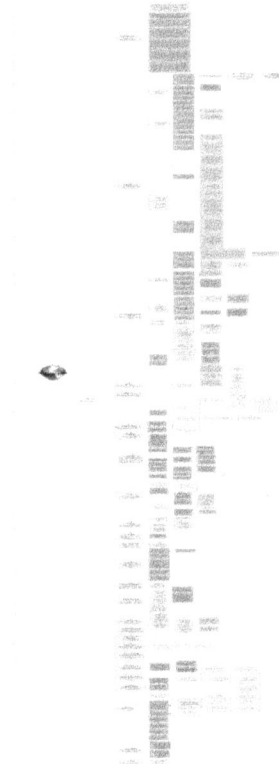

The world doesn't know how close it is to Armageddon. It's just an ordinary day to most. People go to work. People waste time. If they only knew. The risk is great that the story of humanity may soon be, "fought since the dawn of man wielding sticks and stones, invented nuclear weapons, invented social media, destroyed ourselves over nothing, The End."

A wall is approaching the path of humanity. It is our own human nature. It takes us to war every generation, and yet every new generation forgets, thinks they are smarter, that it won't happen again, and repeats it. Next time it will be nuclear. Unless an anomaly stops it. But that anomaly likely has its own challenges.

In every generation, just like how teenagers think they know more than their parents, new ideologies sprout up and say the Jews are evil. It's almost that natural a phenomena.

If Mashiach were here now, he might solve all our problems. But while he is not yet, here are some questions we might want to ask to avoid destroying ourselves until he gets here.

The world is wondering, "why support Israel?" The answer is, because whatever new evil ideologies arise first attack the Jewish people, and unless stopped, attack them next. The Jews are the "canary in the coal mine", or, merely the first target. But how do we explain that to a world that sees each side askew, through a lens of ideological distortion and human emotion? Why is Israel correct and so much of the world's combined viewpoint mistaken? How do we objectively prove it without Israel as the moral compass? The Bible would work, except the Bible is increasingly opposed also. International law would work, except the deciders lose perspective, change definitions, and end up accomplishing the opposite of what it intended. Right now, the same law designed to prevent genocide against Jews, is now being used as a weapon to destroy Israel so the world can again commit....you guessed it...genocide against the Jews.

Whoever and wherever Mashiach is, maybe he is trying to get his email working and his password reset. Maybe he is stuck in traffic, or struggling to get over life's basic obstacles. Someone who will be able to do what he will, will probably have difficulty with the everyday. Moses could not even convince his fellow Jews to want to leave Egypt before demanding Pharaoh let his People go. Per Exodus 4:10, "Moses said to the Lord, "I beseech You, O Lord. I am not a man of words, neither from yesterday nor from the day before yesterday, nor from the time You have spoken to Your servant, for I am heavy of mouth and heavy of tongue.". Per Exodus 6:26-30, "But Moses said before the Lord, "Behold, I am of closed lips; so how will Pharaoh hearken to me?" Moses' brother Aaron had to speak for him. Similar was David, the weakest and youngest brother, a feeble sheepherder, the last person most would have expected to be defeater of Goliath and King of Israel. Maybe we can expect Mashiach, when he comes, to have similar qualities and challenges.

Until he gets here, I'm exploring my ideas to resynchronize the world's moral perspective with Israel and build global peace, like the PeaceMatrix™. We not only need peace, but the right kind of peace; one that protects Israel first as the source of Western civilization's morality, ends global support for Hamas, liberates the Iranian people, and rebalances the world's moral compass.

The system is not fully built, or even complete in its development, but we are close to being out of time.

So I'll show you how it works. Of course, I'll need a problem to show you on.

One theory behind the PeaceMatrix™ is that the world's conflicts are caused by unaddressed questions, and the solutions are in the paths of those same questions.

In the two-year Russia-Ukraine war, a peace deal was a few sentences away from being agreeable nearly two years ago. While peace talks were underway in April of 2022 in Istanbul, former British Prime Minister Boris Johnson flew to Kiev on April 9th, and told Zelensky not to sign the peace agreement that would have resolved the conflict for secure Ukrainian neutrality. Today, we must do more than ask, "what do the parties want?" What the parties want depends on the misaligned power interests of the world continuing to be misaligned, based on more unanswered questions, so the conflict continues. We have a broken system in which the media covers countless attacks, casualties and tear-jerking stories, but gives no attention to the few key missing sentences between Biden's position and Putin's, that may actually resolve the conflict.

In China-Taiwan, maybe unification is right and maybe not, but the world can't even help both the Taiwanese people and China ask key questions about the perfect China-Taiwan relationship, nor calibrate moral compasses, nor properly examine the key question of why the Taiwanese are or are not a unique people and culture, nor foresee the future expansions of China with its expansionist ideological trajectory, nor know the responses of Malaysia, Indonesia, Vietnam, India, Australia, Japan, or the U.S.. Putin misunderstanding the degree to which Ukrainians are a separate people and culture is partly what got us in that mess also, if China wants a road map not to follow into a similar abyss.

The polarization of the Russia-Ukraine war has spread across the globe, aligning Russia, China, Iran and North Korea in a union of each pursuing their own selfish and separate power goals. Once their individual quandaries are deep enough, it's a pretzel we won't be able to untie without nuclear war. Their alignment is helping fuel the conflict between Israel and Hamas, as all of Israel's enemies feel empowered by the backing of Russia and China. Yet, glaring questions like Russia's and China's responsibility likewise go unasked.

Unless we improve our systems, ask and answer the right questions, humanity may well destroy itself for any of countless miscalculations, misunderstandings, or unnecessary quarrels.

It is my dream that the development of a new system of peace-building will reveal that all of the world's conflicts can be solved in a few simple child-like question chains that humanity has simply grown too "adult" to ask. If the solutions to our problems lie where we are too afraid to look, why don't we develop a system to ask the unaskable questions. Maybe answers are just waiting for us to find them.

The goal of the PeaceMatrix™ is to develop the puzzle to find the best questions, by asking questions of questions, until we find the questions that are the keys to unraveling the conflict.

A PeaceMatrix™ dissects 26 categories simultaneously, including:

B-Chain "What is the most important understanding about the parties and terminology for peace?"

C-Chain "What is the most important understanding about the history for peace?"

D-Chain "What is the most important understanding about the parties' wants for peace?"

By developing the question chains, and the patterns they reveal, we eventually arrive at a divine perspective. And we might just solve the whole conflict with a few child-like questions we've grown too adult to ask.

PeaceMatrix™ categories (26) include all the elements of all human conflicts and their solutions, including:

A-Summary and scope

B-Parties and definitions

C-Histories

D-Wants

E-Dispute mapping

F-Communication

G-Unknowns

H-Culture & ideology

I-Writings

J-Morals

K-Rules

L-Unresolved

M-Leaderships

N-Resources

O-Common ground

P-Obstacles

Q-Culture building

R-Communication building

S-Why resolve

T-past successes

U-Unilateral

V-Working backwards

W-Adding parties

X-Solution building

Y-Value Building

Z-Changes

Here are some examples of chain derivative development:

(Please note that the PeaceMatrix™ is a fluid system and questions and coordinates change as the process develops)

(D-Chain - Wants)

D3a - "What is the long-term solution of anyone who wants a ceasefire?"

D1G2 - "How can the Palestinians give up their desire to destroy Israel?"

D3b - "What are the implications of the ideological alignment between those who want to destroy Israel and those who just want to help suffering Palestinians?"

Since the conflict is fueled by the merger of an ideology that wants to destroy Israel with another that wants to merely stop the suffering of the Palestinians and its global implications, why don't we map out their distinctions and develop a solution for one without the other, or even at the expense of the other? Then, the ideological polarization is no longer united against Israel as the out-group. See my article on "The three-Option Plan."

D3b1 - "How do we separate those who want to destroy Israel from those who want to help Palestinians?"

D3b1a - "What peace scenarios involve helping the Palestinians while strengthening Israel's security?"

D3b1a1 - "Who will and won't support such scenarios and why?"

(F-Chain - Communication)

F2a - Why isn't there a communication system for the Palestinians to improve their own political process?

We could ask strategic questions:

(S-Chain - Why Resolve)

S1 - What if October 7th was a distraction so Iran can cause a war, get the bomb, dominate the region, and take control of Saudi Arabia as they have claimed a desire to?

S1a - "How can moderate Arab states work in their own interest to prevent a nuclear Iran?"

S1a1 - "How can the world correct dangerous Russian, Chinese, and North Korean support for Iran?"

S1a1a - "What is the nature of responsibility of these states if Iran fuels conflict in the region?"

We could ask cultural and identity questions:

(B-Chain - Parties and Terms)

B3 - "How are Palestinians separate from or aligned with the broader Arab and Muslim world?"

B3a - "How does Palestinians being Arabs and Muslims affect their ability to negotiate in their own individual interests versus those of the greater Arab cause?"

B3a1 - "What is that broader 'Arab cause?'"

We could map the factual histories:

(C-Chain History and Current Situation)

C1a - "What is the most important understanding of the parties' history for peace?"

Much of the conflict now is because Palestinian ignorance of provable factual history of the Jewish people in the region and its meaning. Maybe there is an education system better than Palestinians teaching their children there was no Jewish presence in the Holy Land before 1948, which obviously leads to conflict.

The PeaceMatrix™ chains work together. If we know the histories, it helps define today's accusatory terms so they turn constructive again:

B5c1 - "Why is the most constructive definition of "colonization" a culture with one homeland and capitol seeking to establish domination over the indigenous people of another land?"

With this definition consistent across all historic colonizers, it is clear the Jewish people are indigenous and have no other homeland, so only Arabs from Arabia who pray toward Mecca five times a day could be "colonizers" in Israel. The U.N. couldn't properly define "colonization" and map out the parties' histories in 75 years. In fact, there's more ignorance and disagreement as to the evidence-based factual history now than when the dispute began. It's like living under communism where the future was certain but the past kept changing. As chain development continues, we can examine why these hypocritical definitional problems exist in the first place.

D1 - "What do Palestinians Want?" leads to D1a - "Why would Palestinians stop hating Israel?"

Arab states have no answer to how the Palestinians would ever develop a Jeffersonian democracy to ensure polarization would be dealt with politically internally instead of all being directed at Israel. After all, no other Arab state in the Middle East is a democracy, so why would a theoretical Palestine be? And if all the polarization is directed at Israel, it doesn't really matter what Israel does or doesn't do, the war to annihilate her will continue. Especially if instigated by Iran. Why would any authoritarian ruler voluntarily take the personal blame for problems when they can just blame Israel?

We can also ask constructive questions, like:

(U-Chain - non zero-sum-game analysis)

U1 - "What potential actions help both sides and harm neither?"

That quickly leads us to ask:

U1b1 - "Why not allow Palestinians who want to leave to relocate (temporarily or permanently)?"

We know this is a key issue, because the first and foremost demand of every Arab state after October 7th was that not a single Palestinian refugee be relocated. If it's the main request of Arab states in a 75-year conflict, as Shakespeare said in Hamlet, "the lady doth protest too much, methinks". That would be the first un-askable question I'd ask. If some Palestinians want to relocate, and are able to relocate, even temporarily, it would help Palestinians, ease pressure from the Iran-induced conflict like letting steam out of a steam engine, and put all parties in a better negotiating position. Regardless whether Israel should or shouldn't pressure Palestinians to relocate, the world still should not force those Palestinians who do want to leave to stay.

So good derivative questions naturally follow:

(X-Chain - Main solution development chain)

X3 - "Why not help Palestinians who want to leave relocate? (temporarily during the conflict or permanently)"

X3c - "What percentage of Palestinians would leave if they could?"

X3a - "Why won't/should/would Egypt, Jordan, and other Arab states accept Palestinian refugees?"

X3a1 - "What would happen if the world compensated countries for accepting Palestinian refugees?"

If Gaza really is an "open-air prison" and "genocide", aren't Egypt, Jordan, and other Arab states, even the whole world that keep Palestinians who want to leave from leaving complicit?

Why does the world force Palestinians to remain there blaming Israel, when the world has no problem gleefully relocating every other people in conflict and taking in millions of refugees?

Israel is not keeping Palestinians in Gaza, it's only keeping them out of Israel. It's the rest of the world keeping the Palestinians in Gaza, by not accepting Palestinian refugees who want to leave.

If this isn't the world's moral compass, why is the world following this policy?

(H-Chain - Cultures and ideologies)

H3 - "Why is or isn't such policy putting the "Palestinian cause" ahead of individual Palestinian's interests?"

H3a - "Why is the world supporting the Palestinian "cause" over the interests of individual Palestinian people?"

H3b - "How is the "Palestinian cause" different than what individual Palestinians want?"

H3c - "Why or why not are "Palestinians" (named after the Philistines which are neither Arab nor Muslim) a unique and independent people versus an arm of an Arab or Islamic ideology?"

And if we keep deriving:

J1 - "Why is or isn't Arab states' refusals to accept Palestinian refugees consistent with the world's morality of helping Palestinians?"

As we continue to ask questions about the Arab ideological phenomenon, we'll get to:

(I-Chain - Writings)

I1 - What are the different Islamic views on following the Quran's statement that God has given the "Promised Land" to the "Children of Israel"? (Qur'an 5:20-21, and 17:104)?

I1a - Why isn't there a growing viewpoint within Islam that supports following this scripture?

(H-Chain - ideologies)

H1 - Why does one Islamic ideology that doesn't follow the Qur'an and caused the Nakba have more support than ideologies that believe the Qur'an's language literally means what it says?

Many in a dominant Arab ideology are not helping Palestinians, because they want to further the "Palestinian cause", even at the expense of the Palestinians themselves. Their view is contrary to their own scripture, and done for ideological expansion purposes. And, the world is supporting it. Nobody disputes that Mecca and Medina are Islamic. Nobody (yet) disputes that Rome is the center of Catholicism. But much of the world is disputing the same for Judaism in Jerusalem. If not Arabs vying to take Jerusalem from the Jewish people, it might be Egyptians, Babylonians, Assyrians, Persians, Romans, Greeks, Byzantines, or other non-indigenous former-conquerors if their ideologies were expanding today.

A question underlying that is:

What is that broader Palestinian or Arab "cause" and why should the rest of the world care?

As I have discussed in my articles on Entitativity, ideologies do not behave according to their proclaimed values in conflict, but like separate living organisms controlling their adherents and seeking to survive, grow, and spread. The Palestinians can't make peace because if they did, the Palestinian "cause" would die, and it doesn't want to die. What else is suicide bombing, if not an expression that the individual adherents are the drones, and the ideology the true living "entity."

With a system more powerful than the U.N., we might fix problems the U.N. can't. We might finally ask the unaskable questions, to uncover the fundamental underlying ideological issues deeper than the U.N.'s structure, international law and definitions, and short-term state policies.

The B-Chain ("Parties") will eventually lead us to examine the implications of Islam, having grown from 90 million in the year 1800 to 200 million in 1900; and from 200 million in 1900 to roughly 2 billion in 2023, and now dominant in 57 nations. By comparison, Jews still have not reached pre-Holocaust levels of just 16 million.

This is the most important thing to understand about Islam as a party.

Egypt used to be a Christian nation. So was Syria. So was Lebanon mostly Christian just 50 years ago, formerly referred to as "the Paris of the Middle East." Afghanistan was Buddhist, as we saw from the beautiful Buddhist statues destroyed by the Taliban. Turkey was largely Christian until the time of the Ottoman conquest, now 98% Muslim. Iraq, Pakistan, and many other nations killed or expelled their Christian, Buddhist, and other ethnic populations. Over a million Jews from Arab countries were expelled and forced to move to Israel. Now, Muslims are expanding to and seeking dominance in not just Israel, but most of the 30 countries of Europe that allow such immigration, but many Asian countries, many African countries, Russia, China, India, Malaysia, Myanmar, Burkina Faso, Nepal, Philippines, and so more. This broader phenomenon wouldn't stop if Israel became a Muslim nation, instead, the exuberance would accelerate it. Within this expansion, Iran's regime has also grown more powerful, without real consequence, to exert influence in much of Middle East.

Yet, with all of that, Israel is being accused of being a "colonizer" and committing "genocide" for being on its indigenous land constituting 0.4% of the Middle East, while Arabs control the other 99.6%.

If Aliens came down to Earth, and looked at the actions of the United Nations, they would think Israel must be conquering the globe, and the Arab world a victim on the verge of extinction. The United Nations is unable to examine both the Israel-Palestinian question and the Israel/world-global Islamic expansion question together, regardless that they are the same fundamental moral question the world's security and future hinge on.

The problem is not people believing different viewpoints, but the ideological polarization over such questions growing into conquest and conflict.If preserving one's culture is natural law that has been moral for 3000+ years if not eternity, and most of the 193 U.N. member states and most other peoples want to preserve their own cultures as well, then Israel is not wrong for wanting to remain Jewish, but rather, an untenable new moral perspective is being ideologically proliferated for the sake of a conquest-motive.

The Palestinians claim they "want" a state, but have rejected proposed states at least five times, when it is not a stepping stone for further conquest of all of Israel.

As we develop the PeaceMatrix™ chains, they present patterns to the heart of the issues, and solutions. And eventually we reach a fundamental underlying question facing the world:

(J- Morals)

J1 - "Why is or isn't Islamic expansion and colonization of the rest of the world's 130 or so remaining non-Muslim countries, and all other religions, tribes, cultures, and peoples consistent with the world's moral perspective?"

How's that for a gem of a question?

In the I-Chain ("Writings") we will see the world has already answered whether a nation's army is allowed to aggressively invade another country and take it over by force in the U. N. Charter's Art. 2 Para. 4 prohibition on use of force as "All members shall refrain…" The U.N. General Assembly also defined state aggression in Res. 3314 (1974), a crime against peace, as "Invasion of a State by the armed forces of another State…"

And in asking what is missing, the world must address the gaping moral loophole left unanswered:

J1a - "Why is or isn't it moral/allowed for one culture to dominate and extinguish another if done by migration, birth rates, bribes, international law, terrorism and/or political pressure?"

That's the gem of a fundamental underlying question the world faces now; that each nation separately, and the world together, must address.

If the world decided that this is moral, or immoral, that would be one thing. But if the world leaves it undecided, there is conflict.

There are many different ideologies within Islam, and many are good, moderate and peaceful people who want to live in peace and balance with all other peoples, including Israel. However, an expansionist ideology within Islam that seeks to (re)conquer Israel is the same conquest ideology seeking to expand and dominate Europe, India, China, Russia, Africa, and countless countries, peoples, and cultures. Even moderate Muslim states and peoples are at risk, with Iranian goals of overthrowing Arab states, and persecution of Sufi Muslims, as examples.

Ironically billions of people today are shortsightedly aligned with the umbrella ideology of their future colonizers. This includes Arab states, who might ask why stopping Iran from going nuclear is a more immediate concern than the Palestinian cause.

Whether Israel can remain Jewish is the same fundamental question as whether every unique and beautiful culture on earth is allowed to remain its own culture. And if they may, then so may Israel. It's natural law. Yet, the world can't answer it, and the United Nations' structure is ill-equipped; one Jewish viewpoint controls one vote, and one Islamic ideology leads across most of 57 Muslim countries' votes. So, they can support "indigenous rights for me, but not for thee." Pure majority rule encourages consumption of smaller indigenous cultures, and thus conflict.

The world's moral compass is skewed, but when clear, the world can no more demand that Palestinians have their own militarized state with a capital in East Jerusalem, than demand that Arab factions in Great Britain or France or Germany or India or Russia or China can have their own state and their own capital in East London or East Paris or East Berlin or East New Delhi or East Moscow or East Beijing with their own armies there. It's the same fundamental moral question. If the world does not prioritize indigenous rights, then what's left is encouraging conquest and colonization. And again, the problem is not Islam, but an ideology within Islam that provides for uncontrolled expansion to, domination of, and elimination of other cultures. At issue are not just derivative questions about nations' immigration or ethnic group policies or international relationships, but also the philosophical, moral, and ideological "why" behind them. Not to mention, it is better for Islam not to allow expansionist factions and ideologies to go unchecked.

The world has answered that colonization is illegal in the United Nations Declaration on the Rights of Indigenous Peoples (2007), for example, Article 3, "Indigenous peoples have the right to self-determination." The world has applied those principles and answered that it is immoral for Europeans to colonize indigenous peoples of South America, Africa,

India, or Islamic countries. Article 6 of the Declaration says, "*Article 6 -* Every indigenous individual has the right to a nationality." The world has already decided it is not "racist" to protect indigenous rights. But the world has not yet applied the same principles and answered, whether it is moral for Islamic peoples to colonize the indigenous peoples of Europe, Asia, Africa, the Americas, and the indigenous Jews in Israel.

The conflicts are the same because the moral question they hinge on is the same.

If you answer the fundamental question first of whether Israel may remain Jewish, consistent with all other peoples' indigenous rights, then its easy to answer the sub-questions about why you can't open up the floodgates with a Palestinian capitol in East Jerusalem, or a Palestinian army, or uncontrolled borders, or right of return. You also answer why other countries must take in Palestinian refugees, and why Iran's leadership is wrong. You also answer why the Palestinians can't be a state, if the ideology is just an arm of a broader "cause" to destroy Israel ahead of the interests of its individuals.

That fundamental question underlies all other sub-questions for policy consideration, because that divide ultimately terminates cultures on the wrong side of it. With this deeper question unaddressed, many nations are allying short-term with Islamic expansionist ideology now, but planning to later conveniently switch when it is their culture at risk. Which, of course, will backfire.

If we ask and answer the fundamental underlying questions, there may be peace, because conquest-minded Islamic factions won't rise to compete over new territory they won't control. And if we don't, conflicts will get worse. Yes. Definitely. For certain.

That, in a nutshell, is how the PeaceMatrix™ works. You find the right question, you solve the puzzle, you solve the war.

In summary, it's not just about moving beyond the moral void of our unexamined, conflict-plagued world currently forcing the Palestinians to remain where many of them don't want to be without a solution, like a forced pawn in a chess game they don't want to play. It's not just about an ideology in the Arab world actually keeping the Palestinians in an "open-air prison".

It's about supporting or not supporting the broader "cause" for which they do so.

And until the nations of the world understand why or why not, they are next.

Shalom.

Daniel Ben Abraham

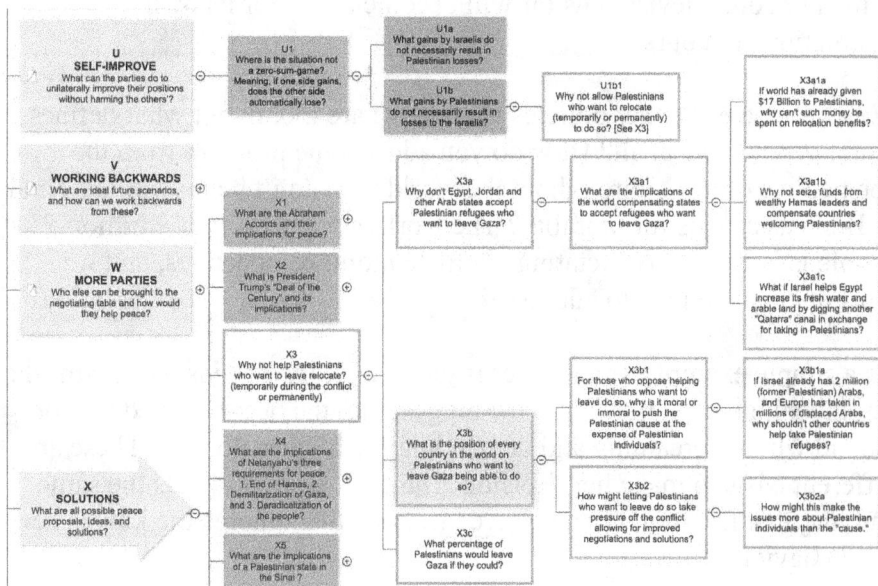

Most recent writings

How the United Nations turns anti-Israel and loses moral perspective

Humanity's best system for preventing war, the United Nations, is an increasingly primitive and deteriorating system, exponentially reflecting man's faults instead of our best. Just as the League of Nations failed to prevent WWII, the United Nations will lead to increasing wars until WWIII, unless replaced or enhanced per my writings, and time is running out.

Why?

Sadly, the system assumes humanity is rational - a fatal flaw. The entire system relies on the hope that nations will negotiate and compromise in their best interests, but it doesn't properly consider human nature. As the United Nations turns anti-Semitic, the best peace-building system humanity could devise thus far will become a tool for its own destruction, and ours.

We as a species further policies we believe are moral, but what defines moral perspective at all? How do you address the problem from the perspective of those throughout the world who don't believe in the Torah or Bible as a basis for morality? The world's population of 9 billion people is increasingly dictating alternate moral perspectives, and we must understand how to address this.

As a simple example, if a mother is giving a group of kids ice cream, she can give the largest portion to the biggest kid, the best-behaved kid, or the kid who is screaming and demanding the biggest portion. These are different basis to make her decision "right". If every kid gets the same amount, that is a moral perspective also, one that ignores the kids' size and behavior.

When we look at the broader world in which Judeo-Christian values don't control, man's perspective of moral righteousness is often an illusion, determined not by an objective standard let alone a constructive standard, but how we feel about each other. In other words, we tend to see our friends as more "right" than our enemies. It's the definition of bias, whether conscious or unconscious. Humanity's inherent tendency to make friends and also enemies - our natural division of the world by in-group versus out-group polarization - infects our reason, applying our emotions to interpretations of the facts to make "our side" appear more justified. It affects our feelings, upon which is based our reason, upon which is based our very formation of language, upon which is based all of international law, upon which is based all its systems. Like a house of cards, it is primitive, because we are primitive.

When that moral perspective becomes anti-Semitic, as the Jews have seen perspectives become time and time again in our 3500+ year history, its systems become worse than useless. They become destructive, and self-destructive.

In the Israeli-Arab dispute, many nations are demanding formation of a "Palestinian state" in Israel, but why?

Many other people are siding with the Palestinians because they are suffering in the current ongoing conflict.

Where does moral perspective lens with which we feel what is right and wrong come from?

First, let's look at who our moral perspective would support if humanity desired to advance to greater peace, prosperity, and achievement. What question may determine who is more righteous and has more rights to the land of Israel?

◦ Who is indigenous? The Jews (Present as a people, culture, nation and religion 3300+ years)
◦ Who has loved the land enough to make it their capitol? The Jews

(King David 1000 B.C.)
◦ Who has better utilized the land? The Jews (Israel is "the Startup nation", the desert now blooms)
◦ Which culture is happier? The Jews (Fourth happiest country)
◦ Which culture is more unique? The Jews (22 other Arab states, including another Palestinian state (Jordan))
◦ Which people have nowhere else to go? The Jews (Arab embargo on refugees is artificial to further the Palestinian "cause")
◦ Which culture may be exterminated without this land? The Jews (proven by 3000+ years of anti-Semitism)
◦ Which group is more of a minority? The Jews (15 million Jews in the world versus 440 million Arabs)
◦ Which group is less a colonizer? The Jews (The only Jewish capitol is Jerusalem, whereas Arabs are from Arabia and pray towards Mecca thus based elsewhere and expanded to colonize Israel)
◦ Which group is more diverse? The Jews (Coming from over 100 nations, and Israel having 2 million+ non-Jews)
◦ Which group has more diverse coexistence? The Jews (All nationalities and religions represented)
◦ Which group is less racist? The Jews (Arabs serve in Israel's government, on courts)
◦ Which group is more democratic? The Jews (Elections and an ideal of rule of law)
◦ Which group is more peace-loving? The Jews (Offered Arabs peace more than five times)
◦ Which group is less genocidal? The Jews (Jews have never expressed desire to destroy 440 million Arabs)
◦ Which group contributes more to humanity? The Jews (Israelis have won 13 Nobel prizes & 213 by Jews, versus only 1 by Palestinians (Arafat) & 16 by all Muslims)
◦ Which group better uses its resources? The Jews (Israel desalinates more water and plants more trees than any nation)
◦ Which group has rights under religious scripture? The Jews (The Torah, Bible and Qur'an all say the Jews are given the land of Israel)

By all the criteria the world claims to value by all the world's dialogue and even the language of the United Nations Charter and writings, the land should belong completely to the Jews in case of conflict. But the United Nations is so ideologically-driven, that if it were tasked with confirming the provable facts underlying these today, it could not admit their truth.

So why is so much of the world trying to forcibly create a Palestinian state? Because the Palestinians:

◦ are suffering, drawing the world's sympathy
◦ were displaced in 1948 because of their own fellow Arabs' attack against Israel
◦ have the political support of an expansionist ideology across 57 Muslim nations, and,
◦ have the support of an extreme, self-destructive, global Leftist ideology

Allowing these factors to form a moral perspective makes the world increasingly anti-Semitic, making Jews the out-group, which skews the world's moral perspective, in a repeating cycle. Much of the world bases its views on sympathy emotions based on short-term casualties, instead of the reasons behind the casualties and conflict, the moral conduct of the parties, and their ultimate aims. Sympathy for the Palestinians triggers serotonin, the bonding neuro-chemical, which overrides the rational analysis of the moral perspectives above. Then, regardless the moral conduct of the parties, the ones suffering more are seen as "right", and deserving of political support.

In other words, almost the whole world is wrong, and Israel is right. Not according to Israel, but according to everyone else's own claimed moral standards.

Could you imagine if, during WW2, when the allies tried to stop Hitler, someone said the amount of innocent civilian German deaths was the determiner of who is right and wrong, not Hitler's plans for the world versus the Allies'?

While many Palestinians are decent people, the majority and especially those who rise to dominate agenda want a state as a stepping stone to destroy Israel. No one desiring a two-state "solution" can withstand debate of why a theoretical Palestinian state wouldn't just build an army to continually start wars against Israel, as we have seen with Gaza's near 20-year independence. Palestinian identity exists not as an independent and unique people, but as Arab Muslims who are an arm of a broader "cause" that the Palestinian people have for 75 years been unable to define.

The world would have endless war if our underlying basis for moral perspective were self-induced suffering and political support caused by ideological motivators. That is our path, if the world insists that the oldest surviving and most contributing nation of Western civilization destroy ourselves for a contradictory and unsustaining moral perspective. Such skewed moral perspectives would destroy Europe just the same next, and then all of Western civilization, and then all other nations, peoples, and cultures on earth.

Again, the Jews must be "a light unto the nations", whether the world is ready to see that light or not.

Dear world, How does it feel to be a Jew-hater?

Dear world,

How does it feel to be a Jew-hater? I'm really curious.

How does it feel to resent the Jews? Does it feel good in your heart?

When you know that Jews can't openly walk down the street in most of the 194 countries in this world, do you get a feeling of satisfaction?

Does history make sense, knowing empires spent their treasures attacking Jerusalem over 20 times over 3000 years to keep Jews out of our homeland, more than any other peoples' capital? Or do you forget such history?

Do you feel joy knowing most other nations oppressed and expelled Jews from almost every place we have been exiled to also?

What about knowing that when we went away like you told us to, the only Jewish safe haven now left is under constant rocket attack - does it help you sleep better at night?

What about when a Jew goes on a dating site, and there are almost no other Jews to date in most of the world's cities because Jews are still fewer worldwide than before the Holocaust, only 15 million out of 8 billion - do you feel love more?

When an extremist ideology seeks to destroy Israel just because it is Jewish, do you side with them, just a little bit, even though you may be next, just because it feels so righteous to persecute the Jews?

How does it feel to see nations going against their own interests just to oppose Israel, and knowing yours is one of them?

Does Europe seem wise to you, having promised "never again", and then waiting barely a generation before replacing the welcomed Nazi Jew hatred with now welcoming Islamist extremist Jew hatred to get rid of even more Jews?

What about when the United Nations spends more time passing resolutions against tiny Israel than all the human rights violators in the entire world put together? Does it feel like truth and justice to you?

What about when the land called Judea for three thousand years is renamed so it doesn't sound Jewish, and is suddenly now called the West Bank? Does it seem factual to you? Do you join in and call it the West Bank also?

What about when 440 million Arabs demand a 23rd Arab state and second Palestinian state so there would be no Jewish homeland, because 99.6% of the Middle East is not enough for them? Does the math seem fair to you?

When a terrorist group attacks villages in Israel and then Israel is taken before the International Court of Justice for trying to get our child hostages freed, do you think "what a fair international legal system we live in"? Do you feel that's the way it should be?

Do you love our world, that promised never again after the Holocaust, but would easily let a second Holocaust happen in Israel, even after October 7th, without the United States?

Or when your country joins against Israel for her strong self-defense response to terrorism? Does it make your heart feel whole?

What about when the world takes in refugees from every other country and conflict zone except Palestinians, just to continue the conflict there and blame the Jews? Does that make sense to you?

Or, how about when almost the entire world demands that Israel divide its capitol and give half to a Jew-hating group that wants Israel destroyed? Does it make you enjoy the security of your capitol city more?

I know, you're not a bad person, right? You just don't believe in a world in which a Jew can walk down the street in his own homeland, hear his own language spoken, see his own people smiling, and smell his delicious food cooking, while you enjoy those very same joys of your own culture, right?

As a Jew, we wonder how it feels to be you. Because, we support the right of every beautiful culture to have their own homeland, including yours, only asking peace in return. Maybe one day that won't be too much to ask. That's our hope. That's literally our national anthem. That's how it feels to be us. In case you wanted to know.

Outsmarting Israel's Conundrum 3.10.2024

In the global chess game, Israel is in check.

The anti-Semites of the world are ecstatic, and ready to sacrifice themselves for the cause of destroying Israel. If you're wondering why the world is so united against Israel, it's a primal reaction like when a shark smells blood in the water. For all our mistakes have come home.

The since October 7th, the forces of anti-Israel aggression have united with the deeply anti-Semitic, the mildly anti-Semitic, the ignorant, and the shared-interested.

Together, they've created a new dynamic. The reason Hamas is not returning the hostages is because they feel they're winning. As long as the Palestinians continue their self-induced suffering, as long as they are stuck in Gaza, as long as Hamas refuses to free the hostages, the world will continue to increase the pressure, restrictions, and demands on Israel, and increasingly limit supplying weapons to Israel. Hamas is ecstatic because it feels this ratchet will continue to tighten the longer this war goes on, until it is strong enough to overrun Israel with Hezbollah, other Arab armies and groups, and Iranian proxies.

If the world were rational, all if would need to do is compare the peaceful, quiet singing and lighting candles of the Jewish gatherings around the world, to the angry screaming and calling for jihad and violence from the Palestinian protests, to know whose side they should be on. But the world is not rational.

The world is ideological. And this is an ideological war.

Israel must learn to address the ideology. We must unlock the key to understanding this, and all wars, now.

Israel is nearly alone against an unholy union of dangerous ideologies. And Netanyahu is fighting Hamas as though killing individual henchmen will win the war, or force them to release the hostages. But it won't.

Israel is not fighting individual Hamas members. Israel is not fighting a person or a group or a country. Israel is fighting an ideology. Hamas is an idea, and you can't destroy an idea by killing its adherents when leaders and funders are in Iran and Qatar. This is why Israel has been fighting Hamas for 35 years. If the Palestinians remain in Gaza, they will provide fertile ground for more Hamas to sprout, especially if they win this war. The problem is not the current Hamas members, but that the totality of the Palestinian problem produces them.

Israel is fighting a collective hive mind called Hamas, embedded in the broader hive mind called the Palestinian cause, embedded in a broader Arab factional ideology, and a number of other interrelated global ideologies.

Hamas will not release hostages because they don't act in their own best interest, but in the interest of the ideological entity. With the heads of the ideology safe in Iran and Qatar, the ideology will happily sacrifice 10,000 Hamas fighters for the "cause", because the ideology is now strengthening with global support from leaders of Iran, Russia, China, most of 57 Muslim states, and increasingly Europe and even the United States.

Israel is in check because Hamas will not release the hostages as long as the ideology is increasing in strength, ideological support, and pressure on Israel, no matter how many Hamas members we kill. Thus, killing Hamas henchman doesn't give you control over the ideology.

The anti-Semitism that arose against Jews in nearly every generation and community we have lived in as a minority, is now going global as our globe becomes a "community." Much of the world will be united against Israel unless we come up with something new, and by the end of this article you'll understand what it is.

First, a recap of the initial failure.

Many things about Israel's defense posture on October 7th were unacceptable. Granted, for whatever reason, over 1000 armed Hamas terrorists were able to cross the border and enter Israel. Israeli intelligence failed. Israeli human intelligence failed. Israeli technology failed. And Israeli border security failed.

But that is just the beginning. Israel needs to be more security-wise at all levels, from bottom to top.

- No more of the liberal mentality that allows hundreds of Israelis to be unarmed and unprotected.

- Israel must never to be so reliant on the U.S. or anyone that she runs out of ammunition or weaponry for the Iron dome in a matter of weeks without U.S. resupplying.

- Israel must always know the exact whereabouts of Hamas leaders, and leaders of other terror groups, whether in Gaza, Iran, Lebanon, Qatar, or elsewhere.

- Israel must never be so focused on one problem as to take her eyes of bigger problems, like Iran's nucleariztion.

- Israel must be ready, willing, and able to carry out strategic strikes at the heart of threats like Iraninan leadership and its nuclear program as immediately as October 8th in response to an event like October 7th. Without that option, we are trapped in an extended distraction, Gaza, while the far bigger threat, a nuclear-armed Iran, fully materializes. Striking Iran would have made Iran look foolish for carrying out October 7th, and made Israel look fearless, stronger, and smarter. it would have prevented the ideological unity we face now, and put pressure on Iran's allies Russia and China for causing the conflict.

- Israel must have a strong policy of deportation of anyone who incites terrorism, and relationships with countries to accept them until they can be relocated elsewhere.

- Israel must have a clear policy that if attacked, she **will** take land in response. When Israel lays out this policy ahead of time, it creates a consequence for attacking Israel, which weakens the anti-Israel ideologies. Israel let the world convince her that taking land if attacked is wrong, when it is moral and necessary.

- Israel should have been able to take advantage of Egypt's economic situation to pressure a Rafah border and Sinai resolution; but instead, the UAE, IMF, and BP rushed to flood Egypt with over $40 billion dollars so Egypt could keep Palestinians suffering in Gaza and the world united against Israel. This despite UAE being part of the Abraham accords, and supposedly an Israeli ally.

- Perhaps worst, Israel failed to fight the brewing ideological war. It allowed Arab states to refuse refugees and keep Palestinians in Gaza, and allowed anti-Semitism to spread on college campuses, in Europe, and global media, increasing the anti-Israel ideology globally until it was so strong it attacked.

- While Israel did not come down against Russia in the Russia-Ukraine conflict, it failed to ensure Russia would not support Iran in its instigation of this conflict, and failed to get Russia (and the U.S.) to isolate Iran once it began.

Enough kvetching. Let's get to the solutions.

Hamas believes it is winning. With the current dynamic, all terrorists need to do is keep the hostages and keep attacking, and they will increase world sympathy and international leverage against Israel.

Let's look at what Israel is purportedly, according to the world, not allowed to do to defend herself.

- Immediately after the attack, Biden urged Netanyahu to delay a military response until U.S. warships could reach the area. Biden sent two aircraft carrier strike groups, the U.S.S. Dwight D. Eisenhower and the U.S.S. Ford Aircraft Carriers into the region in response.

- Then, apparently as an agreed-to condition for U.S. support, Netanyahu agreed not to expand the war outside of Gaza, as Biden did not want a widening this war (also) as he approaches election in November. This means no striking Hezbollah, no striking Iran, no striking Qatar. Netanyahu had to practically twist Biden's arm to retaliate against the attacking Houthis in Yemen.

- On December 23, the Wall Street Journal reported that Biden convinced Netanyahu to halt a preemptive strike against Hezbollah.
- However, despite the Pentagon stating on December 15, 2023 that the Defense Secretary Lloyd Austin ordered the U.S.S. Ford to remain in the Mediterranean, it was announced on January 1st, just two weeks later, that the Ford was returning to port in the United States. Austin made an unannounced in-person visit on the U.S.S. Ford on December 20th. (Not only does the surprise and inexplicable withdrawal send a weak message apparently backing away from Israel, but conflicting messages, which I explain below)

- Arab countries demanded no displacement of a single Palestinian from Gaza, and after Sec. of State Blinken's repeated trips, the U.S. was demanding of Israel the same. Egypt sealed its border, reinforced its fencing, and said any displacement of Palestinians would threaten the Camp David peace deal between Israel and Egypt. Firstly, Egypt has a duty under international law to accept Palestinian refugees who demand asylum at the Rafah border per its own treaty obligations, as I explain in my other article. Egypt cannot blame Israel for this. Secondly, the world gleefully accepts refugees from every other conflict zone, but somehow has a policy of not accepting a single refugee from Gaza in order to place the Palestinian (including terrorist) "cause" ahead of Gaza's who want to leave. Relocating Palestinians during war, even temporarily, should make perfect common sense to the world, who claims they are in an open air prison, that Israel cannot attack Hamas in Rafah with Palestinian civilians around, and that they are starving from lack of aid.

- Purportedly, as part of the first release of hostages deal, Netanyahu promised that he will not attack Hamas members in Qatar.

- Now Biden said in an interview, that there is a red line if Israel invades Rafah. And with a red line, comes consequences. It doesn't matter what Biden meant, but how the world will manipulate Biden politically for saying it.
- Even without having invaded Rafah, many countries are cutting off military aid to Israel. A Netherlands court blocked F35 fighter jet parts to Israel, Belgium supplying gun powder, a Japanese company ended a partnership with an Israeli weapons manufacturer, and Italy and Spain both have embargoes on all weapons. More than 200 members of parliament have committed to banning arms to Israel, including in the U.K, Australia, France, Belgium, Canada, Brazil, Ireland, Turkey, and such efforts can be expected to increase, largely relying on actions in the International Court of Justice.

So, in short, Israel is pretty much not allowed to do anything to defend herself. And as Israel continues to, the world is increasingly united against her.

We must turn this around.

Firstly, Netanyahu needs to understand that it is not just Biden, but likely fiercely anti-Israel Obama who has Biden's ear and is continuing his old disastrous policies. That's likely why we see these 180-degree changes in policy, and Biden saying one thing pro-Israel, but other actions cutting off Israel's options at the same time. Remember that about three quarters of Biden's top 100 staffers are former Obama's people. As anti-Israel as Obama was as president, without accountability, he is even more dangerous.

Next, Israel needs new ideas. It needs to think "outside the box", get creatives, autistics, and geniuses who would never get hired into government, and have them all make a list of ideas. Lock them in a room together, then separately. Then take that list and filter and develop it. Israel will need to implement new ideas faster than the world can unite against them.

Now, addressing the ideological conflict. Israel needs to stop being a pawn on the chess board and start being a player, and a grandmaster.

The way to address an ideological conflict is to control the ideological dynamics. Sure, neutralize Hamas individuals and leaders. But more importantly, understanding how to polarize and repolarize, align and realign, splinter and divide the group ideologies in a conflict is integral to the reducing future warfare and peace building. Ideological momentum shifts for a number of reasons, including the manifesting of a new and more immediate threat, elimination of a key threat, or an opportunity to gain power.

Here's a start.

1. This war happened because Iran's new alliance with Russia and China because of the Ukraine war. Putin gains more from allowing the Middle East conflict to continue because it weakens the U.S. and its support for NATO, so he is siding with Iran. Putin met with Hamas and Palestinian representatives in mid-February 2024, and publicly sides with them. It is not enough to

scowl at Russia in response. Israel must be wiser. Flip it. We must change the dynamic so Putin gains more from alliance with Israel than it does from alliance with Iran and Hamas. One way to do that is to try to resolve the Ukraine-NATO conflict. Israel needs to step up and take a lead role in mediating and negotiating resolutions to the Russia-Ukraine war, now. Israel has over a million Russian speakers, as Putin said once, and is culturally close to both nations, yet neutral enough. Putin wants a resolution and will turn to Israel's favor for it. China doesn't want the risk from Putin's war to transfer onto their Taiwan ambitions, and will also turn to Israel's favor. Place responsibility for this conflict and its resolution on Russia and China, because their alliance with Iran is what emboldened Iran to start this war. Then, isolate Iran. Revive the written deal almost signed in Istanbul, and start asking, "Why?", "How much?" "How many lives?", "Why not?", and the most important, "Who is responsible?" [This disaligns Russia and China from Iran, and realigns them with Israel]

2. Netanyahu needs to understand that Obama, not Biden, is likely pulling the strings, if you haven't noticed the continuation of the disastrous, if not intentionally anti-Russia, pro-Shiite U.S. policy. Biden's weakness is he cares about re-election and defeating Trump most, and Obama cares about being exposed and humiliated the most. Israel needs work with American politics, and involve Trump if Biden does not better support Israel. Tell Trump to demand a FOIA (Freedom of Information Act) for all communications and visits Obama has had with Biden and White House staff. Trump must call out Obama's role in the global chaos. [This realigns Biden with Israel]

3. Start air-dropping supplies for the Palestinians on the other side of Egypt's Rafah border. This will force Egypt to open their border, and make them responsible for not getting aid to the Palestinians at the same time. Israel can say it doesn't want to drop supplies on Palestinians and injure them, and it's better to let Egypt distribute supplies than Hamas. The world can't disagree.

[This aligns the world with Israel and the Palestinians with an exitable Rafah border]

4. Publicly call out world leaders individually asking if they will or will not temporarily take ie: 5000 Palestinian refugees who need food and or medical attention within ie: 24 or 48 hours. Let other leaders say "no" to helping the suffering Palestinians. Then, tell them to at least pay for Egypt to take them. When one country publicly accepts, demand others do the same. This changes Netanyahu's narrative to "we must help the Palestinians", putting pressure wherever he redirects it. [This aligns the world with those who take in Palestinian refugees, letting pressure off the conflict.]

5. File a case with the International Court of Justice that Egypt is required to accept Palestinian refugees that demand entry through the Rafah border into Egypt, per my article laying out the legal basis. Put the humanitarian pressure back on Egypt instead of Egypt putting all the pressure on Israel by closing its border to further the Palestinian "cause." [This aligns the Palestinian cause against Egypt (instead of Israel) unless it takes pressure off the conflict.]

6. Tell Egypt unless it accepts some Palestinian refugees and convinces other Arab states to accept their fair share, Egypt may get them all. Prepare to recognize a Palestinian state in the Sinai if Egypt refuses to cooperate. Suggest that Palestinian state in the Sinai should get the $35 billion in UAE funding going to Egypt's Mediterranean coast anyway, the $8 billion coming from the IMF to Egypt, and $1.5 Billion coming from BP. The investment could save the Palestinians and earn higher profits. Propose no tariffs anywhere in the world for all goods produced and services provided from a Palestinian state in the Sinai to align economic interests. [This aligns the world against this financial transaction unless it goes to help Palestinians relocate] [Israel telling Hamas to release the hostages or she will recognize a Palestinian state in

the Sinai turns Egypt against Hamas.]

7. Have Islamic scholars who believe the Zionist language in the Quran means what it says bring forth a case before the two highest Islamic authorities. Those Quranic texts including, "Remember when Moses said to his people: 'My people, remember Allah's favour upon you when He raised Prophets amongst you and appointed you rulers, and granted to you what He had not granted to anyone else in the world." (Qur'an 5:20). "My people! Enter the holy land which Allah has ordained for you; and do not turn back for then you will turn about losers." (Qur'an 5:21) "And thereafter We [Allah] said to the Children of Israel: 'Dwell securely in the Promised Land" (Qur'an 17:104). Demand legal opinions from the two highest Islamic authorities on **all** the reasons why this language supposedly doesn't mean what it says. Say Israel wants to respect their religion and not force the Palestinians to do something unIslamic. The more scholarly thought and debate provoked, the better. Propose economic developments in Southern Lebanon, Western Jordan, and the Sinai, because such will be consistent with scripture, and thus prosperous. [This supports emergence of a zionist Arab ideology to counter the Arab anti-Israel ideology.]

8. Israel must align with countries and peoples with similar moral positions in parallel challenges. For example, put out a dossier explaining the similarities between the Kashi Vishwanath Hindu Temple dispute in Varanasi, India, with the Gyanvapi mosque that was build beside it. In that case, there is a parallel moral question of a mosque being built atop another religion's holy site during an Islamic conquest, (like the Jewish temple), and then the indigenous culture regaining control. It's in both Israel's and Hindu India's interests for the same moral code to apply. Both India and Israel can agree on a global moral rule that an indigenous holy site or territory later conquered belongs to the indigenous culture that was there first. Thus, Israel stands as a bulwark for the same principle as India. If Israel is reconquered

as Islamic extremists seek, they will demand the same reversion in India, and elsewhere. [This aligns Israel's moral perspective with PM Modi, a Hindu revival, and conservatives in a nation of a billion people. Other nations will join in support for this moral perspective as a basis for international indigenous cultural rights.] [Similarly, the Rohingya Muslims in Myanmar, who are being relocated, and conservative political parties in Europe, who seek to limit Islamic migration, political expansion, and increasing cultural dominance.]

9. Know exactly how to help a democratic revolution in Iran. The Iranian people deserve their freedom, and justice for Iran's leadership attacking Israel should be Israel helping the "free Iran" movement to turn Iran into an ally. [This disaligns the Iranian leadership further from its people, and aligns the Iranian people with Israel.]

10. Israel is wrong to rely solely on the West, and should improve alliances amongst like-minded Asian countries, including India, Japan, China, and other Southeast Asian nations. Many of these countries have similar cultural concerns regarding Islamic expansion and have much stronger limitations on this than Europe. Both China and India are removing Mosques, and India is passing a citizenship law that excludes Muslims. After India broke off Pakistan to be a separate Muslim state, it now regardless has 200 million Muslims and growing in India in conflict with its Hindu Revival. Asian countries not based on Abrahamic religions tend to be less anti-Semitic. [This will align Israel's global moral perspective with many states and billions of people]

It would be nice to have more friends.

Netanyahu should better grasp the true war

Dear Prime Minister Netanyahu,

Thank you for your valiant efforts so far in this great struggle. While your determination to free the hostages is admirable, we must always remember that we are Jews. We accomplish far more with brains, and Gods' help, than we do with brute force.

The reason Hamas is rejecting hostage deals is because they feel they are winning. It doesn't matter that Israel is morally right, if the world doesn't see it, and unites against us. As I have explained, one cannot destroy an ideology by killing its low-level henchmen. Israel's 35-year-long fight against Hamas proves that.

Let's use our God-given brains to get out of this mess. This isn't a wrestling match, it's a chess game. And I should remind you that most of the world's chess masters have been Jewish.

Firstly, if you're playing a chess game and a pawn attacks you, you don't necessarily devote all your forces to killing that pawn. You go after their king. That's how you win the game. October 7th is a small fraction of what will happen if Iran goes nuclear.

On October 8th, Israel should have been ready to attack Iran's nuclear sites, its leadership, and Hamas leaders in Qatar. The conventional war would have been harder, but ideologically the entire world would not be uniting against Israel for month after month of persecuting the world's favorite victim, the Palestinians. Ideological lines would have been drawn differently, and Russia and China would have been be under global pressure for supporting Iran.

Now, Israel is largely alone, and still doesn't seem to understand how to fight an ideological war. You're not fighting the leaders of most of 193 countries all mysteriously united against Israel. You're fighting one ideology that is growing in power. That one ideology controls the thinking of many world leaders. That ideology can be divided, but you must understand how.

Here are some ideas for re-diverting the ideology now united against Israel, as examples:

1. Now that the International Court of Justice in The Hague has said Israel needs to open up more access to allow in humanitarian aid into Gaza, Israel should seek the same Court ruling for Egypt. The Court can't take a contradictory position regarding Egypt and maintain its objectivity. Israel should ask the world for MORE aid for Palestinians, and ask it be sent to Egypt with tracking devices to ensure Egyptians are getting the aid across the Rafah border to the Palestinians. This opens Egypt's Rafah border, and also puts the onus on Egypt for not getting enough aid to Palestinians. No Israeli official should mention aid without mentioning efforts to get it through Egypt in the same sentence. Stop dropping food aid in Gaza and drop it in the Sinai so Egypt can deliver it. Let the news media focus on Egypt's border policies and actions. [This changes the ideological conflict from the world being united against Israel, to being between Egypt and the Palestinians, and Egypt and the other nations]

2. As Egypt is a signatory to the convention on refugees, it must allow Palestinians seeking asylum to cross at Rafah into Egypt until their status can be determined. Egypt is a signatory to the United Nations Convention Relating to the Status of Refugees, adopted in 1951, with accession on May 22, 1981. At the Rafah Border, Egypt is bound by the principle of *"non-refoulement"* as defined in Article 33 of the 1951 Convention. Article 33 of the Convention, entitled *"Prohibition of expulsion or return refoulement"* states, *"No Contracting State shall expel or return ("refouler") a refugee in any manner whatsoever to the frontiers of territories where his life or freedom would be threatened on account of his race, religion, nationality, membership of a particular social group or political opinion."* The French word

"refoulement" refers to *"faire reculer"*, which in refugee law *"a la fois l'üloignement du territoire et la non-admission a l'entrüe."* (See D. Alland and C. Teitgen-Colly, *Traitü du droit d'asile* (Paris: Presses Universitaires de France, 2002), 229) This translated, is *"both removal from the territory and non-admission to entry"*. Thus, Egypt under Article 33 of the 1951 Convention is prohibited from denying admission to entry to Palestinians who show up at the Rafah border seeking asylum. Egypt cannot blame Israel for its own obligations under international law. [This makes the ideological conflict between Egypt and the Palestinians, and Egypt and the other nations] See my article in the Times of Israel for more: https://blogs.timesofisrael.com/egypt-must-accept-refugees-at-rafah-per-intl-law/

3. Call for a temporary humanitarian relocation of the Palestinians from Rafah. Tell the world that even Israel will take in 5000 Palestinians refugees who want to leave Gaza and come into Israel, but, if and only if, the other nations each take in on average the same amount. 193 nations x 5000 is 965,000. Let the nations of the world say no. [This makes the ideological conflict between the Palestinians who wish to leave and all the other nations]

4. Blame Qatar for this war continuing. With a GPD over $200 billion and the world's richest country per capita, you should demand Qatar's wealth, condos, and fancy cars be given to the Palestinians as compensation. [This turns the ideological conflict from the world against Israel, to Palestinians against Qatar's leadership, and Qatar's leadership against its own population.] Threaten to file a case in the International Court of Justice against Qatar on behalf of and for compensating the Palestinians. Reserve the right to strike Hamas members in Qatar after Ramadan unless Qatar arrests and hands over all Hamas leaders.

5. Announce that southern Lebanon was part of ancient Israel, and if Lebanon doesn't stop Hezbollah from attacking and disarm them,

Israel will annex southern Lebanon. [This makes the ideological conflict between Lebanon and Hezbolla] A main reason these conflicts keep happening is because Israel doesn't keep land when it wins, and thus fails to create conflicting goals between the terrorist group and the underlying population.

6. Tell the UAE and IMF that instead of investing $50 billion in Egyptian resorts west of Alexandria, that money should go to help the Palestinians as it is enough to build a Palestinian state in the Sinai. Put pressure on the UAE that the funds be put to better use, and invest in a Palestinian state instead to kill two birds with one stone. Calling for a Palestinian state in the Sinai, per my other writings, gives everyone demanding a Palestinian state what they want, except being a stepping stone to destroying Israel. It takes the conflict our of a zero-sum-game equation. [The ideological conflict becomes those who want to help Palestinians with statehood and peace and the Mediterranean coast is fine with them, versus those who only want a Palestinian state only to better attack Israel with] It is no more absurd than 140 countries recognizing a Palestinian state on Israeli territory. https://danielbenabraham.substack.com/p/recognize-a-palestinian-state-how

7. Ultimately, you are seeing a repeat of the disastrous, anti-Israel, pro-Iran, anti-Russia foreign policy of Obama, who left Iran enriched (no pun intended) and three endless American wars ongoing. A Democrat re-election would mean this present course worsening and quite possibly the end of Israel. Most American Jews do not understand that is happening, and believe the facade that Biden is helping Israel. They don't understand the Obama-Biden anti-Israel U.N. resolutions fuel threats against Israel. You must communicate this to the American people with the key election approaching.

8. This war was started by Iran emboldened by Russian and Chinese support, and those ties are the key to resolving it. Lead a

delegation to try to negotiate a resolution to the Russia-Ukraine war. Helping resolve that dispute would align Israel's goals with Russia and shift the poles of this entire conflict. [This would take the ideological conflict from Russia, China, Iran, Hamas, and Hezbollah against Israel, to Russia aligned with Israel rather than Iran because Israel is the key to ending Russia's war]

All of these I have explained in detail in my other writings. Our fundamental mistake as a species in misunderstanding wars is believing leaders are responsible. You didn't "decide" to go to war any more than Sinwar "decided" to go to war. Rather, the Hamas/Iran/Russia/China ideology got so strong that the individual Hamas members couldn't resist attacking. Soldiers don't cause wars. Leaders don't cause wars. Living, growing ideologies with their own separate interests control both like puppet-masters. And my writings explore the keys to understanding them.

FAQ on Israel and global security 4.8.2024

The following is a question and answer on PeaceMatrix™ Entitativity theory applied to the Arab-Israeli conflict, anti-Semitism, and current global security challenges.

1.   Why does the world only like Jews when they are victims?

Empathy triggers brain neurohormones like oxytocin, activated in the central nervous system with in-group identification, bonding, and trust, and which conversely is reduced when dealing with an out-group. When Jews are suffering, people identify with Jews and the resentment of anti-Semitism is overcome. The challenge is for the world to like and identify with Jews without Jewish suffering as a prerequisite. Also, conversely, increasing anti-Semitism instead of countering it, media coverage of suffering Palestinians triggers similar empathy, and causes people to be emotionally conditioned to side with Palestinians on a neurochemistry level more powerful than logical analysis of the moral character of the parties in conflict. The world forgets Palestinian suffering is self-induced by their hatred of and attempts to destroy Israel because it is Jewish and prosperous.

2.   Why did anti-Semitism increase after October 7th before Israel even responded?

The savage brutality of the attack triggered primal human responses like pack mentality, spiking polarization directed at Jews as the out-group. Those who dislike Jews saw Israel appear weak, triggering an urge to attack a wounded enemy like a shark feeding frenzy. The gathering of large pro-Palestine crowds globally is not merely "speech"; but allows for connection with the ideological collective hive mind through contagious group neurochemistry.

3.   Why is terrorism increasing globally?

Our international legal system based on responsibility of nation-states allows non-state ideological groups to avoid responsibility while gaining power, and while helping supporting states carry out agendas. The world's Muslim population has grown from 90 million in the year 1900 to nearly two billion today, now dominating in 57 nations. Sympathy for Palestinians has increased acceptance of terrorist methods, exponentially increased by the political and economic pull of 57 nations often representing one Islamic "nation" ideologically. The expansion of the broader religion causes ideological subsets to compete for power gains in new prospective territory not under an established Islamic authority's control.

4.   Why are Israel's enemies ganging up on her?

In Arab culture, there is no word for "compromise" and peace is historically typically sought only temporarily by a party that is weaker. That's partly why there are no Arab democracies. New support from Russia and China empowered Iran and Hamas with new superpower backing, while international law is increasingly used as a tool against Israel. Hamas triggered a primal, ideological response akin to a feeding frenzy seeing Israel as the common prey. With the deeper ideological polarization, Israel's enemies are emboldened and more willing to sacrifice themselves to further their larger collective ideology's "cause".

5.   Why does humanity have war?

Wars are not caused by soldiers, individuals, or even leaders usually. Nor are wars typically over land or resources or the claimed excuses. Wars are caused by ideas that come to life and possess masses of people turning them irrational in group ideological phenomenon. This "ideologies are separate puppet master entities of human behavior in conflict" idea is the subject of my PeaceMatrix™ Entitativity theory. Wars happen because humanity encounters moral questions it can't answer, and polarizes around them for political gain of the separate factions. Such ideological phenomena sideline rational thinking in their human adherents, gaining power from increasingly polarizing their in-group adherents against another out-group, causing a reciprocating series of errors of judgment on both sides. Adherents on both sides believe they are each defending themselves, and aggressive leaders break down the systems that maintain peace (treaties, rules, norms, laws, interpretation of laws, language, communication, even self-control) until the only option left is war. To a lesser or greater degree, the leaders we blame for war are merely the key pawns of the ideological entity. My PeaceMatrix™ theoretical world peace building system is a global communication system to help peoples resolve the world's unanswered moral questions.

6.   Why do ideologies turn anti-Semitic and spread?

With Judeo-Christian values as the source of Western civilization's moral perspective, for new ideologies to rise, they must differentiate themselves. As they increasingly gain followers in number and ideological adhesion, they turn evil by definition by diverging from source values. The ultimate differentiation from natural law is to turn anti-Semitic, broadening the personal subconscious unacceptance of Jewish people across an ideological narrative. New and growing ideologies unite with even contradictory ideologies for the combined power gains of doing so, which are a more powerful influence on adherents than their claimed values.

7.   Why is international law (the United Nations Security Council, International Court of Justice, Human Rights Commission) turning anti-Semitic, more critical of Israel than all other nations and genocidal leaders put together?

The U.N. is an imperfect system of rules that degrades over time as the shared values and interests evolve without a system to reset them to their source other than war. In short, nations lose their moral compass over time as ideologies morph and shape over time. Anti-Semitism is not rational, but primitive emotion emotional, which in turn controls ideology, which in turn controls moral perspective, which in turn controls language, which in turn controls legal interpretation. 57 Islamic nations led by one ideology that polarizes against Israel, increasingly turning perspective away from the fact that opposing Israel because it is Jewish is the cause of the conflict and contrary to the world's purported morality. In a nutshell, the world's morality is largely built on Judaism, and yet a contrary moral perspective holds up to 57 seats while Israel cannot even have a seat at the Security Council. As the United Nations votes, for example, that a 3000 year-old Jewish Temple does not belong to the Jews, it becomes morally corrupt and destroys its credibility until it renders itself ineffective. (Similarly the U.N. is unable to prevent war between Permanent Security Council members who hold great sway and veto power, this polarizing the world between them.) because of such factors, the U.N. will increasingly degrade until nuclear war.

8.   How do liberalism and Leftism lead to conflict?

Liberalism is merely open-mindedness to new ideas, responsible for much of mankind's progress. It is a natural divergence from norms, requiring conservatives to balance its diverging moral compasses. This is why balanced news media and open debate is so important. Without this balance, liberalism becomes open-minded to even the idea that the establishment ideologies that allowed it to flourish are evil, and thus skews toward Leftism. New moralities form to compete with the establishment's, for example, censorship is okay if one is offended, etc. Leftism is thus the tribalism of human individuals innate self-destructive tendency rationalized across a collective ideology. Unmoderated, the new, undefined and even contradictory ideologies form a collective psyche that controls the masses and eventually destroys the society by accepting leadership from any dangerous, polarizing source promising the entity more power. Meaning together, it seeks the destruction of humanity.

9.   What's wrong with overcoming or overthrowing the establishment power structure?

Nothing. The problem is the desire to also overthrow alongside it for mere power gain the establishment's moral compass which built our modern civilization on codified national law and brought peace countless times against naziism, communism, and authoritarianism. Thus, only a force seeking to reestablish such ideologies can try to overthrow natural law.

10.   Why is the world becoming more anti-semitic?

Communities have turned against Jews wherever Jews have lived for over 3000 years; the first 1000 in conquests of the Holy Land and the next 2000 in nearly every community Jews have been exiled to. As the world becomes a community, the same phenomenon of unrelated and contradictory complaints unite under the subconscious umbrella of anti-Semitism for the political gain the directed irrational hatred provides.

11.   Will anti-Semitism get better?

No, it will only get worse. As the world becomes more intertwined and complex, nations will find increasing reasons to unite irrationally against the Jewish nation. Since the reasons given are mere excuses for primal phenomenon with ideological support, no rational solutions (land, Israel's treatment of the Palestinians, Palestinians statehood) will suffice while Palestinian ideology insists on acting in their own disinterest, and external influences remain. The emotional divisions will grow deeper into the collective psyche controlling what appears to man to be reason, while true rational analysis of the conflict's moral questions will increasingly be sidelined. The world will need to do this as alternate ideologies rise to challenge the foundational moral perspective of civilization, the Jews'.

Answers to #12 - #25 are available for full member newsletter subscribers.

   11. What about the Abraham Accords, peace treaties with now six
       Arab countries, and possibly soon the Saudis?
   12. Are Arabs immoral relative to Jews?

13. Why are many Arabs opposed to Israel's existence?

14. Why is the Israeli-Palestinian-Arab conflict so difficult to resolve?

15. Is a two-state solution possible, and if not, why?

16. Why doesn't the world understand this?

17. Why don't Arab neighbors realize they would be next?

18. Why should non-Muslim states support Israel?

19. Why should Muslims support Israel?

20. Can Hamas be destroyed in the current conflict?

21. What solutions exist then?

22. How can we end or reduce anti-Semitism?

23. How do we get out of the immediate conflict?

24. If the key is ending the Russia Ukraine conflict, why is that war happening?

25. How would you resolve the conflict(s)?

Why Iran Can't War - Iran's difficulty waging war in metaphysical context

What's more powerful, a bomb, or the ability to turn the world from anti-Israel to anti-Iran in one day?

You may have noticed a complete 180-degree change in the war since the killing of Iranian General and Quds Force Commander Reza Zahedi on April 1, 2024.

The ongoing war went, in one day, from virtually the whole world united against Israel, from U.N. action and talk of increasing arms embargoes, to after Zahedi's removal, our Arab allies, including Saudi Arabia, Jordan, and others, working with the United States and European countries to unprecedented cooperation defending Israel against a common enemy, Iran.

If only Israel understood how to redraw such conflict lines intentionally. Such futuristic strategy would be a new level understanding of both warfare and peace-building, able to change the landscape of all wars to come. It just so happens to be the subject of a new theory on conflict reorganization strategy and resolution strategy I am developing called PeaceMatrix™ Entitativity theory, but I digress.

But this move is more than Israel merely showing its capacity to deliver justice for October 7th which Zahedi allegedly was responsible for, and prevent future attacks. The world went from calling for cease-fires and arms restrictions on Israel, including even the United States allowing a U.N. Security Council Resolution calling for an "immediate ceasefire" on March 25th, to an entirely different landscape in this conflict. After 35 years of fighting Hamas, Israel has hit a nerve untouched since its removal of Qasem Soleimani in 2020, which reportedly even Netanyahu backed out of at the last minute allowing Trump to handle alone.

Now, Iran threatens Israel, if she should retaliate to Iran's retaliation of 300 missiles and drones fired this past Shabbat. And of course, Israel should. Why?

- With all of Iran's size, its size is its own disadvantage. Iran is required to expend great time and resources moving any troops and military assets thousands of miles in its own territory, let alone to reach Israel. Israel can take advantage of this and continually force Iran to move its resources around.

- Iran appeared incredibly weak, with half of its projectiles not even reaching Israel from their own failures.

- Iran's entire economy is based on energy production and transportation, and is extremely vulnerable to even unsophisticated attack and sabotage.

- Iran's currency the Rial has hit a new low plunging 30% of its value as of late March, to 613,500 to the dollar. https://apnews.com/article/iran-currency-rial-record-low-economy-2c59af5dfa9bbbb9e5286352e2899cf5

- Iran's population is already discontent with its economy, wasted resources on its nuclear program, international sanctions, and obstacles to democratic reform. After an attempted democratic revolution that was thwarted y force in 2009, Iran is still increasing crackdown on outspoken dissidents, like those following the death of 22-year-old Mahsa Amini in custody of Iran's morality police. https://www.hrw.org/news/2023/09/15/iran-crackdown-dissent-ahead-protest-anniversary Iran's government has also been criticized for its increased executions recently, executing 853 people in 2023, allegedly on drug possession, but suspicions are in efforts to quell an uprising of dissent. https://www.amnesty.org/en/latest/news/2024/04/iran-executes-853-people-in-eight-year-high-amid-relentless-repression-and-renewed-war-on-drugs/

- Russia needs Iran to help Russia in its conflict with Ukraine, as Iran has been supplying Russia with drones and other resources, and as a key member of BRICS international economic cooperative. But a growing conflict in the Middle East turns Iran from an asset of Russia to a liability. All those nations siding with Russia that are opposed to Iran's nuclearization could be re-aligned in the new polarization that would result from increased tensions involving Iran, harming Russia's access to military assets, and its ability to circumvent global sanctions. Putin called for restraint to avoid a catastrophic clash in this matter. https://www.reuters.com/world/middle-east/putin-tells-middle-east-pull-back-catastrophic-clash-2024-04-16/

- China, which already tiptoes through foreign policy as though blindfolded across an icy road with great uncertainty of its correct path, stands to ostracize more nations by supporting yet another nation waging yet another escalating war, with its economic if not military support of Iran. China, which has threatened yet even another action of its own regarding Taiwan uniting South Asian countries in opposition, likewise called for restraint.

- Not to mention, ISIS-K and its agents spread across Syria, Iraq, Afghanistan, and Pakistan, who have carried out attacks at Solemani's memorial service inside Iran, as well as the terrorist attack at a theater in Moscow, and may look to take advantage of greater instability.

Law professor Alan Dershowitz wrote an article appearing in the New York Sun about how Israel had a moral and legal right and responsibility to stop Iran's nuclear program, and how this escalating series of attacks may provide a golden opportunity (paraphrasing). https://www.nysun.com/article/why-is-biden-stopping-israel-from-attacking-irans-nuclear-weapons-program

Israel attacking Iran's nuclear program in response to Iran's attack would make Iran's attack last week appear foolish, encouraging even more of the Iranian regime's supporters to abandon it.

So now, Iran is caught in a catch-22. To fail to answer Israel's escalating responses, whether against nuclear facilities or elsewhere, grows discontent with its Mullahs, whose entire ideological political infrastructure (not to mention its alliance with its proxies like Hezbollah and Hamas) is based on hatred of Israel. Yet, continuing the reciprocating escalations weakens Iran's leadership both vis-a-vis its own internal ideological counterforces, and its key international support in Russia and China which underlies the entire Gaza conflict.

As I've written elsewhere, war and peace are ultimately not about how many bombs you drop on low-level henchmen or even leaders, but understanding and guiding the ideologies behind them.

Maybe Israel is now finally playing chess.

The Beauty of Zionism

When I was growing up, I didn't want to be Jewish. Now, I think Zionism is one of the most beautiful things in the world.

Not that I even knew what it even meant though, growing up. I was born under communism, where Judaism wasn't allowed. The only reason I was given a Jewish name is because Daniel was on the list of approved Christian names. My parents were too afraid to tell me I am Jewish, so I never knew until we escaped and fled to America. My grandparents didn't tell my mother she is Jewish until she was 16 to protect her.

I don't blame them. Most of my family died either in concentration camps, taken to labor camps, in the Budapest Ghetto, shot, drowned, or otherwise killed for being Jewish. Both my parents were an only child, and I grew up not even really knowing what it's like to have cousins. So if you think you understand the current conflict, just show me an Arab family in any of the 22 Arab countries spanning the Middle East who don't know cousins and are afraid to tell their own children who they are.

Throughout my life, I never fit in. I was always different, in my head, awkward, introverted, quiet, nerdy, and painfully logical in a social world. I did everything conceivable to make friends, but rarely could. And good friends, hardly ever. I was just different. I was bullied a lot, and being openly Jewish was just one more reason to be picked on, so I avoided it all.

One day, in my late 40s, after lifetime of lacking meaningful connections, friends, the right partner, and even without properly understanding myself despite endless self-examination, I found myself in Israel. For the first time in my life, I felt home. After escaping from communism and never feeling home, after growing up on the streets of New York and never feeling home, after living in California never feeling home, I felt at home in Israel. I saw faces that inexplicably looked familiar, people's inherent nature felt like my own, and a lifetime of not fitting in finally made sense.

I felt like everyone there is another person like me. I felt like I understood others and they understood me. For the first time in my life, I felt home. I had never walked down the street before fully feeling surrounded by "my people", despite loving America all my life with all my heart, and being a full-blooded patriotic American. America is my beloved nation and home in a sense, but amongst Jews I realized we share a heart and soul like I never imagined I would find. Maybe it was how some American Christians might feel when they are not only in America in their community, but go to their home church. It was like living over 40 years, and moving nearly 20 times, and then finally finding long lost family – a whole tribe of them. It was one of the most beautiful things I've ever experienced in my life.

I couldn't understand why the world was so hell-bent on not allowing Jews to have a safe-haven. Every culture in the world has a home where they can enjoy their own food, listen to their own music, walk down the street surrounded by people who speak the same language, who have the same history, religion, and culture. Why was the world trying to keep Jews from enjoying some thing without which I never felt fully human? Why was "Zionism", simply the idea that the Jewish people should also have their own tiny homeland like every other people, so offensive to the world? It's certainly not about land. The Arabs have the other 99.6% of the land in the Middle East, and five Arab countries simultaneously attacked Israel one day after its founding on May 14, 1948 to try to destroy Zionism. An evil ideology has maintained that goal ever since, harming all reasonable parties including many Arabs and Muslims in the process. It existed before the formation of modern Israel, when the ruling Ottomans had laws against Jews being able to buy land, and it will continue until the world finally sees the light of reason.

Palestinians can walk down the street in 22 other Arab nations including the "Palestinian Mandate" State of Jordan with similar language, culture, and religion. They can walk down so many streets and smell their delicious food cooking, see familiar faces, and hear their cultural music. Why were so many Arabs against Israel's very existence? Israel is just 0.4% of the land in the Middle East, and doesn't take away anything from them, but adds to the culture and economy of the region. And why is so much of the world on their side? Why has empire after civilization after nation attacked Jerusalem, and after many Jews were expelled by the Romans, attacked Jews in virtually every other community we have lived? And when we want to leave our persecution and go back to our own tiny little home, why do they attack us there worst of all? Why is the world trying to deny us what every other culture enjoys? Why did the Roman Emperor rename the land "Syria-Palestina" after Israel's long-gone enemies? And why did the world recently rename Biblical Judea and Samaria the "West Bank"? If they merely wanted another Arab state alongside Israel, they could have declared one when the "West Bank" was controlled by Jordan and Gaza was controlled by Egypt between 1948 and 1967.

I mean, I have an idea why. Over 3500 years ago Jews were amazing engineers and builders, so the Egyptians enslaved us, and had us build their cities. They probably had their excuses then too. From 3000 years ago in Jerusalem, we were a center of civilization. After bringing the world "thou shout not kill" and "thou shalt not steal", empire after empire spent blood and treasure to take Jerusalem from the Jews, and tried to conquer it more than any other city in world history, over 20 conquests. For the next 2000 years after that, nation after nation we fled to grew to resent us. Why? Because, for example, our hand-washing rituals which date back to Leviticus – they got plagues, and did not understand why we didn't, countless centuries before doctors invented microscopes and discovered bacteria. And, they hated us for it. A study by the National Institutes of Health suggests Jews may have a gene mutation which helped protect against plagues. Perhaps we knew that since Egypt, but is that our fault? We are overrepresented in Nobel Prize winners over 100-fold, and apparently that's really disturbing to some too. Is the fact that we have a culture of learning our fault? Today, the world runs on an international legal system that focuses more on criticizing Israel than all other nations put together, and finds the idea unbearable that the Jews again can be right, and much of the rest of the world wrong. Today, not only Judaism, but Christianity, America, Europe, and Western Civilization and many others are under attack.

After all, if Jews can have a tiny homeland free from conquest, so can every other unique and beautiful people and culture in the world. And that apparently doesn't fit fell in some people's plans.

What are we to do, besides survive, and be who we are? What can we do besides help the world to see the light and figure things out, the same way the world realized the Nazis were wrong, and the pogroms were wrong, and our enslavement in Egypt was wrong, and killing us in countless localities where we have lived was wrong. So that's what we're doing now. We're trying to stay alive, while helping the world to realize that burning babies in their cribs is wrong, and kidnapping women and children and elderly is wrong, and wanting to deny us a home is wrong, and 9/11 was wrong, and conquering us and other peoples and cultures is wrong, and so on. Just imagine if such acts would become the world's morality, and you understand what the Jewish people stand in the way of, even if we must stand alone. Those doing these things are wrong. And all those who side with them are wrong, whether they shout with ak-47s or speak artfully in the media or at the United Nations. Their combined goal is the same. No more Zionism. No more Israel. No more safe haven for Jews. And with that, no more Jews. And then you are next. And the closer they get to that goal, and the closer they get to war, the further any middle ground becomes. And the further a chance at reason becomes. And soon the only question left is, whose side are you on?

The True Reason We're Not Getting the Hostages Back

Who is winning this war?

The IDF said they killed over 12,000 Hamas members. That sure sounds like Israel is winning, no?

Despite these numbers, after months of no new hostage deals, on March 25th, Israel recalled its negotiators from Qatar saying Hamas is not interested in making a deal, not even 20 hostages for 1000 Palestinian prisoners. American negotiators reached the same conclusions - that Hamas does not want any deal that does not let them claim victory, survive, and reconstitute.

If Israel were winning, Hamas would quickly make a deal to release the hostages.

If Hamas gained from such a deal, Hamas would quickly make a deal to release the hostages.

But they're not.

Hamas is not releasing the hostages because they believe *they* are winning.

They believe they are gaining by *not* releasing the hostages.

The strategic problem is, Israel's definition of "winning" and Hamas' definition of "winning" are different.

Why the discrepancy? What don't we understand?

Israel spent 35 years fighting Hamas, like Hezbollah, because even removing their top leaderships is only temporary. As soon as the fighting is done in Gaza, Hamas will declare victory and reconstitute from the Palestinian population. The U.S. similarly failed in Afghanistan, not understanding the Taliban ideology is more deeply embedded in the Afghan psyche than Jeffersonian democracy. American generals mistakenly viewed the Afghan War like World War II, where we fought to regain territory in Europe, and that territory was wrongly equated with progress and victory.

Israel hoped to turn the Palestinian people against Hamas, but it largely didn't work. According to a poll released March 20, 2024 by the Palestinian Center for Policy and Survey Research, 70% of Palestinians are satisfied with the role Hamas has played during the war, and in an election, Hamas would today defeat Fatah 2 to 1.

Our fundamental misunderstanding is the belief that we are fighting an army. That is wrong. Hamas is *not* an army. Hamas is not a group. Hamas is not a person. Hamas is not even every single Hamas member. Hamas is not even the leadership.

Hamas is an idea.

The reason we are not getting the hostages back is because our true enemy is not Hamas fighters or leaders, but the ideology, which is a separate entity from these, happy to sacrifice human followers or "adherents", both Hamas members and Palestinians, in order to continue growing the broader ideology and its support globally.

While Palestinians are suffering and Hamas is losing members, the support for the Hamas-led Palestinian "cause" and the broader ideology is growing globally. Translation: Individual Hamas members are losing, but their ideology is winning.

And if the ideology is winning from the conflict continuing, they won't release the hostages.

Then why did Hamas make the first deal? Because the first deal apparently came with assurances from Netanyahu that Israel would not attack Hamas leadership in Qatar. Netanyahu should have put a 60-day time limit on that promise. He gave up being able to fully pressure Qatar, demand Qatar arrest Hamas, or hold the Qatari government complicit. Those benefits are why Qatar stepped in to negotiate the first hostage release, not out of the kindness of their hearts.

Dear Golda Meir was right in saying, "If we have to have a choice between being dead and pitied, and being alive with a bad image, we'd rather be alive and have the bad image."

But soon, that bad image will unite the world against Israel.

Now, 143 nations in the General Assembly just voted to recognize a Palestinian state, so it can bring in more armies and weapons to better attack Israel with. Even the United States is increasingly joining the global descent into anti-Semitic lunacy by cutting off weapons shipments to Israel, and increasingly opposing Israel in the United Nations. To reiterate, the world wants to welcome into the community of "peace-loving nations" [U.N. Charter Article 4] a group led by Hamas, Hezbollah, Fatah, and Iran, that by 72% supports October 7th, and largely wants genocide of all Israelis. And by the way, that U.N. draft resolution called for a "contiguous" Palestinian state.

The anti-Israel ideology is gaining globally as protests in the streets and college campuses signal the future. Meanwhile, Israel fights a war with her hands tied behind her back. Supposedly, Israel is not allowed to attack Hamas leaders in Qatar, nor hold Qatar responsible for hosting Hamas nor for over a billion dollars sent to American universities for Hamas ideological support. Supposedly, Israel must give Palestinians food and water and medical care during the war, and is responsible for their safety. Israel obeys the Arabs' prohibition against allowing Palestinians to leave permanently or even temporarily, or even opening the border with Egypt. Israel also does not annex land used to attack her from in Lebanon or Gaza. Israel faces an increasingly global weapons embargo, accepts the restrictions on attacking Lebanon, and on only attacking Iran symbolically.

In the international sphere, the International Court of Justice, the United Nations General Assembly, the Security Council, the Human Rights Commission, and the International Criminal Court are all hearing arguments against Israel, and even the U.S. is increasingly opposing Israel in the U.N..

To Hamas, this is not a war of territory or soldiers lost. (Of course, as Israel promised not to annex Gaza.) This is a war for the world's moral perspective changing from thinking Hamas are the "bad guys" to thinking Israel is the "bad guy".

The world's moral compass, the very lens through which the world sees what is right and what is wrong, is increasingly skewed to see Israel as the "bad guy". So why wouldn't Hamas continue? Soon, to them, international law will require an arms embargo and isolation of Israel, and recognition and arming of a "Palestinian state," so the real war can begin.

Netanyahu complains that "the world is ganging up on us", as if it's inexplicable, unstoppable, mysterious anti-Semitism.

Jewish teachings that anti-Semitism is "irrational" are correct, but that doesn't mean there's nothing we can do. Anti-Semitism is irrational in the sense that anti-Semitism does not benefit the lives of its human adherents infected with it. In fact, they will suffer just to have Jews suffer. Anti-Semitism is also irrational in the sense that most people protesting on American college campuses are illogical, many unable to even name which river or which sea they are screaming about.

But "irrational" doesn't mean working by *no* rules - just a set of rules we don't understand.

Imagine we understood those rules.

We try to convince the Palestinians they have a better life in peace like the U.S. tried to convince the Afghan people they would have a better life without the Taliban, but it doesn't work. We try to convince the world it is being anti-Semitic, but it doesn't work. Because, logic doesn't work on people who are ideologically-driven. The ideology spreads like a virus, and you can't very well rationalize with a person to not be infected with a virus.

Hamas is an idea, and the only thing that can defeat an idea is a better idea.

But what?

The growing global ideologies fighting Israel work by a set of rules Israel does not understand, but must learn.

Dr. Jordan B. Peterson talks about Swiss psychiatrist Dr. Carl Jung as having said, "People don't have ideas. Ideas possess people."

Hamas is an ideological mindset hosted in the collective psyche of its members and supporters, but which supersedes and transcends any, and even every member. You cannot destroy Hamas by killing its members. Not even all its members, but certainly not its low-level members. That's why the Jews were not just required by God to destroy all the Amalek, but all *memory* of them. If the IDF took out every Hamas member, Hamas' memory would still live on in the Palestinian mindset, and reconstitute.

Wars are not caused by individuals. Wars are not caused by soldiers. Wars are not even caused by leaders. Wars are caused by ideological hive minds that live in the subconscious collectives of their adherents. And it makes sense, as primitive man faced most mortal threats as a group. These ideologies commandeer the rational thinking of their adherents, make each side believe they are morally right and defending themselves from a bigger threat in the other, increasingly polarize one in-group against another out-group, and break down the systems that maintain peace until war is increasingly the only option left.

What if our best experts' thinking thus far is two-dimensional? What if there is a vastly-superior, even divine understanding of human nature, terrorism, group psychology, global anti-Semitism, and how to alter their course, just waiting for us to grasp and apply it?

First to know, logic and reason can't convince most idealogues. Such ideologies don't spread randomly nor by sneezing like viruses. There are very specific reasons, rules, or "dynamics", by which ideologies grow, spread, change or mutate, control the rational thinking of their human adherents, turn anti-Semitic, unite with even contradictory ideologies, and compel their human adherents to unite against an "out-group" like Israel, and attack, whether with rockets and bombs, or legally in the U.N., diplomatically, or politically. This is a force that far transcends even our nation-state system, so we can make peace with a nation's leadership, but not its people, as 94% of Arabs still do not recognize Israel.

Second, ideologies do not behave according to their proclaimed values and goals. This is why they align with parties with completely contradictory values, and want more whenever they receive what they previously claimed to want.

Third, ideologies have interests separate from their adherents, even ALL their adherents, especially as they approach conflict. Ideologies will have adherents to do things in the disinterest of the individual, and even the whole group. The growing ideology will act against the interests of Palestinians, all Arabs, and even all Islam eventually.

Fourth, ideologies have interests superior to their adherents. Ideologies will compel adherents to behave as to override their own individual interests.

Fifth, ideologies supersede the rational thinking of their adherents. The driving force is a primal, emotional compulsion through the amygdala infinitely stronger than reasoning of the pre-frontal cortex. And, which will look and feel like reason and wisdom and morality to its adherents.

Sixth, the behavior of an ideology is organized, and has structure.

In other words, separate and superior interests to their adherents overriding rational thought in sum translates to... ideologies behave less like a belief system or set of rules, and more like a living organism, with a host/parasite relationship over their adherents. I call these understandings and their progeny "PeaceMatrix™ living organism" theory and "PeaceMatrix™ Entitavitity Theory", and I have been researching and developing them for years.

Understanding the ideology is a separate and independent invisible party to the dispute, means it must be addressed separately from all the other parties to the conflict. An analogy I use is, instead of looking at the pieces on the chess board, now you see the real player sitting at the table.

This understanding is a next-level game-changer for the first time in human history in understanding how to defeat every terrorist and anti-semitic nation, group, nation, and ideology growing worldwide, now, and forever.

I call such ideologies the puppet-masters of humanity's wars. They seek to grow, spread, survive, and become more powerful like any other living organism. Ideologies in conflict are the superior and sovereign being, and individual human adherents are the mindless drones sacrificing their lives to strengthen the ideology. In fact, what is suicide bombing, if not the ideology saying *it* is the sovereign entity, and the individual adherent is just the drone sacrificing his life to strengthen the collective hive mind's grip on the emotional psyche of the adherents.

Thus, knowing how to neutralize the ideology is completely different from, and far more important than neutralizing the individuals.

Hamas can't make peace, because then the ideology would die, and it doesn't want to die. It behaves as a living organism controlling the subconscious collective minds of its adherents. The ideology strengthens itself from compelling individual Hamas members to war and the global ramifications, so Hamas causes war.

Of course, it makes perfect sense now. Anti-Semitism didn't increase after October 7th by 1000% before Israel even responded because people rationally thought the terrorist cause was moral. It increased because the ideologies opposed to Israel appeared more powerful. The biased perspective that Israel is morally or legally the "bad guy" is only purported justification in hindsight of the primal subconscious alliance with what they feel is the stronger ideology. South Africa isn't bringing Israel before the International Court of Justice because of an objective moral position, but compelled by the power gain of joining the anti-Israel ideological collective.

If the global political, legal, and media landscape are such that the broader ideology gains from more conflict with Israel, Israel will be attacked again.

But when we understand the dynamics of such ideologies, we can change their course on a macro-level, and their human adherents will change their behavior to follow suit. Then, we will finally be dealing with the chess player sitting at the table, not the pieces on the board distracting us. Then, we will be able to stop wars strategically, like how we metaphorically fight fires by cutting off fuel sources and oxygen instead of one bucket of water at a time.

Imagine Qatar were turned against Hamas. Imagine Russia turned against Iran, Iran against Hamas, Egypt against Hamas, and even the Palestinians genuinely against Hamas. It's a far superior way of neutralizing conflicts, rather than bombing low-level Hamas henchmen while the leaders are safe and the ideology is strengthening.

If Israel understood how to actually destroy the Hamas ideology, they would release the hostages overnight.

The understanding of the dynamics of ideologies in conflict is the beginning of peace for Israel, and all mankind.

And that understanding is coming.

(End of Part 1)

(Part 2 starts here)

So, ideologies come to life as collective hive mind ideologies, and take over the rational thinking of their adherents.

Such ideologies have interests a) separate from their adherents, and b) superior to the adherents. In other words, the ideology functions as the supreme being, and the individuals who follow it are just the drones, like a bee protecting the hive.

The key in this is ideologically-led people do not follow the claimed ideals of their ideology. They do not act rationally to achieve its claimed goals. After all, the ideology has been promoting to help Palestinians for 75 years, and has not done so. This is because the ideology compels human adherents to act in their disinterest, that helps the ideology grow, spread, and deepen in the emotional psyches of human adherents.

What is suicide bombing, if not the ideology saying that it, not the individual human, is the sovereign entity? The sovereign being sends humans to die to strengthen the ideology's grip on the population. So in fact, tragic Palestinian suffering increases the strength of the ideology. My research and development of what I call "PeaceMatrix™ Entitativity theory" explores understandings of the dynamics of ideologies and how to shape, direct, weaken, and even theoretically destroy them.

Not only do ideologues not follow the claimed values of the ideology, but they don't know that they don't.

And neither do those fighting against an ideology.

An ideology behaves like its own separate simple living organism desiring to spread, grow, and strengthen. Deeply embedded in a population's emotional psyche, the ideology will sacrifice all the interests of its adherents, including their leaders, in order to keep growing, spreading, and opening its emotional grip on adherents.

Fighting the adherents does not bring victory, as we see from this 75-year conflict.

All of our current Western counter-terrorist strategy is misguided because we do not treat the ideology as a separate entity, a separate player at the poker table. We humanize in our analysis the ideology down to the interests of the individual, and that is a mistake.

Israel is killing individual adherents of the ideology, but not harming, threatening, or destroying the ideology itself. The ideology is actually gaining power, growing and spreading, in both number of adherents and in their level of adhesion to the group mindset globally.

The war in 1948 was an ideology using Arab armies to try to destroy Israel. Now, the Arab leaders pander to both sides, while the anti-Israel ideology grips 94% of their populations, spreads across the globe, and uses international law and diplomatic pressure to disarm Israel, to ultimately carry out the same goal.

As a side note, there is a trope used to dissuade killing terrorists which says killing terrorists only creates more terrorists. That's not what I am saying at all. Israel has a right to defend herself and kill every terrorist she can. I'm saying that without having deeper understanding, that still won't be enough alone. But as a strategic matter, when you kill individual members of a deeply-embedded ideology, the ideological grip on the subconscious of its other adherents is indeed strengthened. As ideologies strengthen, adherents become less rational, less self-interested, and even more willing to sacrifice themselves for the ideology's benefit.

Hamas is an idea, and the only way to defeat it is with better ideas.

The only thing that works, is addressing the ideology. And I don't mean primitively trying to convince people not to support their ideology. Of course, that's like trying to convince a living being to give up its own power and commit suicide. Why would it?

But if we can understand how to divide its adherents against one another, unite opposing ideologies against it, and otherwise ideologically weaken it, we can affect the landscape of conflicts on a macro level, instead of cutting off a lizard's tail only to have it grow back for decades on end while the true "war" grows.

If the Palestinians had a vote tomorrow, and voted in new leadership in Gaza, and vowed to kill every Hamas member on sight as anti-Palestinian and anti-Islamic by a large majority, and the world decided with unity to not do anything that might help Hamas, and if Iran was isolated, and every Palestinian completely rejected any outside interference materially or ideologically, Hamas would be finished as an ideology.

But that is not what is happening. The university campus protests, the nations refusing to sell arms to Israel, and the United Nations' and world's combined actions help strengthen the ideology of Hamas and the ideology of the "Palestinian cause."

If you want to know something scary, what is happening in large pro-Hamas protests and college campuses isn't "speech", it is individuals connecting with the collective hive mind contagious neurochemistry of the ideology, same as hitler spoke to crowds to brainwash people. If you weren't brainwashed by his speech, your brain caught the contagious neurochemistry from those standing next to you.

On Israel's current path, killing all of Hamas and getting back many hostages is far less likely than a symbolic victory now.

So how do you fight an idea?

Read my other writings.

Some ideas:

Simplified, the "Palestinian cause" is actually two ideologies. One wants to live in peace with Israel, and the second wants to destroy Israel. The cause of the entire conflict is that they are intertwined, so when the world tries to help one, it helps the second. If you divide them perfectly, you resolve the conflict.

Israel can always unilaterally recognize a Palestinian state in the Sinai.

Israel can take control of the Rafah border. Egypt must take in Palestinian refugees who wish to voluntarily seek asylum at the Rafah border per Egypt's responsibilities under international law. (https://blogs.timesofisrael.com/egypt-must-accept-refugees-at-rafah-per-intl-law/)

Israel could pressure the UAE to funnel their $35 billion land development investment for West of Alexandria there.

These would separate Hamas' ideological goals from those of the moderate Palestinians, and the world's goal of helping the Palestinians from Hamas' goal of destroying Israel.

It would no longer be a zero-sum-game conflict.)

Starting with the United States taking Palestinian refugees, Israel should offer to take 10,000 Gazan refugees but only if every other nation takes in an average of 10,000 refugees. (10,000 x 193 = 1,920,000)

Israel might tell Lebanon that unless they disarm Hezbollah, Israel will annex southern Lebanon, or recognize a Palestinian state in southern Lebanon. (That would turn Lebanon against Hezbollah, and Hezbollah against Hamas.)

Qatar should pay for hosting Hamas and causing all this Palestinian suffering. Why aren't we demanding Qatar's multi-million-dollar condos go to Palestinians? (That would turn Palestinians and the whole world against Qatar.)

Israel gave up the opportunity to unite the world alongside her and against Iran with a more serious attack on its nuclear program, weapons manufacturing, and oil facilities, and instead went with a mere symbolic attack that sent Iran to Pakistan and North Korea, three guesses why. (Escalating conflict with Iran would unite Sunni Arab allies with Israel's own goals, and make Iran a liability to Russia instead of an asset. Perhaps you didn't notice Putin's meetings with Hamas and Iran whenever either of them so much as sneeze.)

Israel could facilitate the union and cooperation of anti-Mullah Kurds in Iran, anti-Mullah Arabs in Iran, and other democratic and rogue forces in Iran with Arab allies. (That could unite ideologies opposed to the Mullahs, increasing their combined power towards bringing democracy to Iran.)

Israel could help resolve the Russia-Ukraine war and become more valuable to Russia than Iran, which would help Putin resolve this whole conflict by its root cause. (That would divide Russia's interests from Iran's and help Russia reduce the world's conflicts in everyone's best interests, including America's. I'd honestly love negotiating this one myself.)

A Palestinian state elsewhere is a key category of solution. And if you offer to recognize such a state (or more than one), you turn everyone opposed to it against each other, instead of against Israel. You don't need to expel a single Palestinian, as many will voluntarily leave once they have the option. As I say, it's not Israel keeping Palestinians in an open-air prison. Israel is only keeping them out of Israel. It's the rest of the world keeping Palestinians in an open air prison with its intentional global embargo on accepting Palestinian refugees seeking asylum. Once Palestinians have the option to voluntarily leave, then Hamas's interests will be at direct opposite to those of the Palestinians. Because then, the more aggressive Hamas gets, the more Palestinians will leave, the weaker Hamas will get, as Hamas can't survive ideologically without the supportive population.

A capitol in East Jerusalem, which has never been an Islamic capitol, is as silly as a new Islamic state in Russia with East Moscow as its capitol, or a new Islamic state in China with East Beijing as its capitol, or new Islamic states in France or the U.K. or India with East Paris and East London and East New Delhi as their capitols. In other words, most of the world should be on Israel's side, but won't be until we correctly understand this conflict.

Do these ideas sound crazy to you? That's funny, because fighting Hamas for 35 years while the world grows increasingly anti-Israel sounds crazy to me.

War in a Box

The danger in letting ambiguous U.S. and world interests direct the Gaza war:

Rabbi Lord Sacks wrote, *"it is only by being what we uniquely are that we contribute to humankind what we alone can give."* And, our Jewish teachings say that we are not and should not be like the other nations, if we wish to be a light unto the nations.

A key to war strategy is, if you do nothing else, be unpredictable. But Israel is not allowed. Israel is fighting a war reliant on U.S. and global bureaucrats and politicians with a flurry of interests separate from Israel's, and this mistake will cost Israeli lives and security. And the greatest danger in letting ambiguous at best U.S. and global interests direct Israel's war in Gaza may be letting go of our own good judgment.

Israel is now fighting a war in which she is not allowed to win. Which makes sense, led by the United States which didn't win in Vietnam, didn't stop Bin Laden before 9/11, didn't win in Afghanistan in 20 years, didn't win in Iraq in 18 years, and is now in a two-year escalating stalemate with the Russia-Ukraine war, also funding both sides of that war. The U.S. runs wars not to win, but to become endless government programs passed on from one administration to the next, and Israel fell into the same trap.

Israel can't fight endless wars. And with better strategy, Israel shouldn't have to.

It's easy to see how Netanyahu was led into this trap. Apparently as an agreed-to condition for U.S. support, Netanyahu agreed not to expand the war outside of Gaza, as Biden does not want yet another widening war as he approaches the election in November. But it was still a mistake to agree. Failure to hold Iran and Qatar responsible on October 8th will lead to more wars against Israel until our apparent predictability and the enemy's perceived safety are undone. If Israel cannot fight a long-term war, whether for supply or political reasons, it must be reading, willing, and able to resolve wars quickly to deter the next one. Israel failed. Israel is fighting an "inside the box" war, and things can't get better until that changes.

On December 23, the Wall Street Journal reported that Biden convinced Netanyahu to halt a preemptive strike against Hezbollah. And of course, that conflict is only escalating as a result.

The war was sabotaged from the start, when Arab countries immediately demanded no displacement of a single Palestinian from Gaza. After speaking with Secretary of State Blinken, the United States blindly adopted this disastrous position as policy. If Israel can't move Palestinians out of the way, Israel can't fight a war without looking evil in the media. The global Palestinian refugee embargo is what's really keeping them in an open-air prison, not Israel. And of course, if the Palestinian population can't vote with their feet and leave voluntarily, Hamas only gets more powerful the more war Hamas causes. Whereas, if Palestinians could voluntarily leave, Hamas would be harming the "cause" by causing conflict, and clashing with their own ideology. But it's not allowed, right? The result is Hamas would still win in an election held today, and will begin to reconstitute the moment the military campaign ends.

Israel's winning move from the beginning has been to allow Palestinians to voluntarily transfer, if they so choose, to the Sinai, temporarily or otherwise, to third countries or otherwise. but Israel filed to act according to her morals. If Egypt would not accept refugees at the Rafah border, Israel could have declared Egypt responsible for the humanitarian crisis, demanded all food be sent to the Sinai for Egypt to distribute, and forced open the border. Israel could have brought a case before the International Court of Justice for Egypt failing its obligations as a signatory to the Convention Relating to the Status of Refugees, and resolved the war before the trial. See: https://blogs.timesofisrael.com/egypt-must-accept-refugees-at-rafah-per-intl-law/ Egypt threatened withdrawal from Camp David, but Israel holds all the negotiating cards with the ability to allow more voluntary transfer of Gazans if Egypt doesn't cooperate. Israel should be viewed as committing a humanitarian act leading Gazans to safety and food across the border, and Egypt trying to keep the Palestinians starving in a war zone as the "atrocity." By the way, why is there no dramatic video of Palestinians banging on the Rafah border wall begging for food and access, as should be on every screen in the world.

Israel dropped the ball, like on this potential deal to resettle Palestinians in the Sinai back in 2014. https://www.timesofisrael.com/lawmakers-back-reported-bid-to-settle-palestinians-in-sinai/ Abbas doesn't have to agree. Recognizing a provisional Palestinian state in the Sinai would split the ideology in half. It would bring Palestinians everything that want, except closer to destroying Israel, giving Palestinians a choice between happiness and Hamas. Then, the more Hamas fights, the more Palestinians transfer to the Sinai and set up a thriving state as an alternate option. But again, we're not allowed, right? There's nothing immoral about giving Palestinians an additional state in the Sinai. If Hamas is such bad leadership that they force many Palestinians there, that's not Israel's fault. The point is, if Palestinians have a choice, to either leave or have a state in the Sinai, Israel isn't as good a scapegoat - that's why such a second option is opposed by the Arabs. Point being, if Israel is going to recognize a Palestinian state, it should do so in the Sinai and see how that works out before betting the farm.

Why does the rest of the world help enforce the "open air prison"? Israel simply hasn't been morally strong enough on giving the Palestinians a choice of freedom. That's what "Free Palestine" should mean. Israel is afraid of being criticized for dislocating Palestinians, but if it's their choice, Israel is the good guy for setting them free.

You know what else weakens Hamas ideologically? Annexing land they use to launch attacks, because then, there's actually a political consequence to attacking Israel. But again, Israel is not allowed.

Then, Israel stood by and did nothing as the UAE, an Abraham Accord ally, invested $35 billion to bail out Egypt with land development West of Alexandria, just to avoid the correct use of those funds, enough to build an entire Palestinian state in the Sinai.

The next limitation was, that reportedly as part of the first release of hostages deal, Netanyahu promised that he will not attack Hamas members in Qatar. Netanyahu should have put a 60-day time limit on that promise, but he didn't. Qatar and Egypt are not negotiating hostage deals out of the kindness of their hearts. They are keeping Israel just close enough to prevent Israel from doing what is needed. Had Israel demanded Qatar arrest Hamas leaders or we would hold Qatar responsible, from its leaders to its oil production, Israel would have been in a win-win situation.

Now, many countries are cutting off military aid to Israel, like a modern day Judenboykott from 1930's Germany. A Netherlands court blocked F35 fighter jet parts to Israel. Belgium is not supplying gun powder. A Japanese company ended a partnership with an Israeli weapons manufacturer. And Italy and Spain both have embargoes on all weapons. More than 200 members of parliament have committed to banning arms to Israel, including in the U.K, Australia, France, Belgium, Canada, Brazil, Ireland, Turkey. Such efforts will only increase.

The next tragic limitation was Israel succumbing to U.S. and ally pressure, and only carrying out a symbolic attack on Iran. As a Passover miracle, the United States and many of the Arabs came to Israel's side when Iran attacked. All the parties in all the wars realigned like a game of musical chairs. Iran's weakness is that an escalating war with Israel would turn Iran into a liability instead of an asset for Putin, which Putin won't tolerate. Iran knew it made a severe mistake in attacking Israel, by giving Israel full justification to severely hamper its nuclear program, oil, and weapons production in response, among other targets. The Iranian people could have had their freedom. Even President Erdogan of Turkey complained that Israel was escaping criticism on Gaza during that time. But instead, Israel said no to stopping the real threat of Iran's nuclear program through the resulting chess game, and decided to go back to focusing on the distraction, the Palestinians. The Mullahs must have cheered before running to meet with Pakistani and North Korean officials days later, three guesses why.

The U.S. sends an aircraft carrier one day, and brings it home days later.

The U.S. helps defend Israel against Iran one day, and frees up $10 Billion in funds to Iran the next. In fact, after an announcement of $6 Billion is when October 7th happened.

The U.S. says its relationship with Israel is "iron-clad" one day, and days later abstains from a U.N. resolution calling for a ceasefire that doesn't condemn Hamas, and stops ammunition shipments to Israel in the middle of a war to get hostages back, giving terrorists hope to continue fighting.

If Satan were running things, trying to encourage and fuel both sides into escalating war, he couldn't be doing a much better job. Or, it could just be Biden's moral and intellectual weakness - a riveting combination producing a similar result.

Such a two-faced U.S. Administration is incidentally classic Obama, who secretly pushed for and abstained from U.N. Security Council Resolution 2334. That resolution is the reason much of the world looks at Israel as an illegal occupier today. Fiercely anti-Israel Obama speaks to Biden regularly, speaks directly to the White House Chief of Staff, and up to 75% of White House staff has been his former staff. Obama hated Putin, Obama hated Netanyahu, had a mysteriously pro-Shiite Administration contrary to longstanding U.S. pro-Saudi policy (not to mention, walked without Secret Service with the Palestinian Authority in Ramallah.) Such policies became strangely familiar again from day one of the current Administration. So if there's anything worse than Obama, it's Obama policies with no accountability.

Most Jewish Americans still have no understanding of how truly duplicitously the U.S. is juggling Israeli and anti-Israeli interests or why, or which side the coin will ultimately land on as the conflict worsens.

Mysteriously, the actions that Israel needs to take to win, are the very same actions the anti-Israel forces are pressuring the U.S. to pressure Israel not to take.

If Israel is not being faithful to her values, and God's law, and her military, and her hostages, and her good judgment, and long-term security in order to artificially suppress the true conflict to help the Biden / Obama team get re-elected before things really fall apart, Israel should consider what four more years of such wars will do for Israel's chances of survival.

Message of Weakness

Why the symbolic bombing of Iran was a mistake:

In short, if we are unpredictable, we are strong. If we are predictable, we are weak, and wars go on forever.

After over a thousand Israelis were killed, raped, and kidnapped, I am deeply concerned about the mere "message" (of weakness) sent instead of stopping Iran's nuclear program or giving the Iranian people their freedom.

This symbolic gesture by Israel comes after Iran-sponsored terrorists six months later still refuse to release hostages because they feel they are winning, and after Iran attacked Israel with 300 missiles and drones, and after Iran threatens to go nuclear and wipe Israel off the map. I feel this way not because I want revenge, but because weakness based on liberal fantasy invites more aggression and war. And I say this as a former liberal, speaking about "liberal-minded" defense policies over politics, though they suspiciously often coincide.

We've become so liberal-minded in Western democracies in our understanding of foreign threats that we've invented new rules for war, whereby our enemies have the will and seek the means to destroy us, and we send polite messages back that we have the means but not the will to stop them from doing so. And we think we're safe while real threats grow.

Or perhaps you didn't notice, that Western democracies haven't won a war since adopting this new mindset. The U.S. went liberal-minded, was unsuccessful in Vietnam, allowed Bin Laden to escape and commit 9-11, was unsuccessful in Iraq in 18 years, was unsuccessful in Afghanistan in 20 years, and now insists on telling Israel how to conduct war. And Israel allowed herself to be in this position. If Israel is incapable of fighting long-term wars, isn't it then necessary for her survival to be ready, willing, and able to remove the leadership of threats so they won't try? Or at least be sufficiently unpredictable?

With this current mentality, after Hitler killed 6 million Jews, we might have arrested him, given him a firm lecture as a warning, and then set him free sternly waiving our index finger with the words, "if you do that again, next time there will be a hefty fine and maybe even jail time."

Liberal-minded Europe has so forsaken its own defenses, from Islamic extremists and Putin, that it's not a question of IF Europe will end, but which welcomed threat will annihilate Europe first. And yet they criticized American cowboyism from under the very blanket of freedom we provide, from Naziism, Communism, and now Russia. Meanwhile, they still buy energy from Russia and allow the U.S. to fund the endless and escalating stalemate in Ukraine. By the way, did you forget when Putin was firing at Ukraine's nuclear power plant just for spite?

Here's the problem. Those in the dream world of liberal-minded defense wake up from their beautifully-optimistic if not dangerously-rosy fantasy and turn conservative only when reality threatens their existence....which is often too late. Seeking peace is always beautiful until it brings worse war. In one study, three times more survivors of the 9/11 World Trade Center attacks were reported to be more conservative after, as opposed to more liberal. https://www.jstor.org/stable/41262946 A study published on the National Institute of Health (NIH) website says after the trauma, many became, and stayed, more Republican. https://www.ncbi.nlm.nih.gov/pmc/articles/PMC3876262/ In short, threats turn us more conservative-minded, and when we believe (wrongly) that such horrors of history will never happen again, that human nature magically changed, we become more liberal. Hence the adage and Irving Kristol quote, "a conservative is a liberal who's been mugged by reality."

This liberal mentality of "nothing is going to happen" has taken over much of Israel. Just the same, it took over the Noa festival, with no armed guards, and even Israeli citizens reportedly prevented from having firearms.

We need to seek peace. But in the interim until our peace systems improve, we need to stop nuclear proliferation and aggression from leaderships like Iran's. This was one opportunity to do so. To understand why it would have worked, read the post Why Iran Can't War.

What's the Big Deal About Rafah

Why is the entire world in an uproar about Rafah? Why have the Arab states so strongly opposed an IDF operation there? Why do they strongly pressure the U.S. to oppose the IDF operation in Rafah? The Arab states strangely haven't been so insistent on anything post Oct. 7 since they told U.S. Secretary of State Blinken that not a single Palestinian is to leave Gaza.

Seems suspicious, no?

As Shakespeare said in Hamlet, "The lady doth protest too much, methinks." Meaning, when so much of the Arab world is so strongly against something, maybe there's something good for Israel in it.

The entire conflict rests upon Palestinians being trapped in Gaza, unable to leave, like a pressure cooker.

Why? There is no political process in Gaza, so Palestinians have no option but to support Hamas, even if Hamas brings suffering. And so they support Hamas. And Hamas can continue the war endlessly.

But if the Rafah border with Egypt were opened by the IDF, then, Gazans can vote with their feet, and leave. It's like letting steam out of the entire pressure cooker conflict.

Mystery solved, our Arab allies aren't afraid IDF will kill civilians, but that Israel will open the Rafah border to Egypt's Sinai.

What might happen if Israel opens the Rafah border and lets through Palestinians who voluntarily wish to leave and seek asylum in the Sinai in Egypt, even temporarily?

Let's see if we can count the game-changers...

- Hamas actually becomes accountable, because if they cause war and Palestinians leave into the Sinai, Hamas' ideology will be weakened (extremists are like fish who can only swim in a sea of moderates).
- Hamas has to return the hostages, because if not, Palestinians will flood out of Gaza as the conflict continues.
- The Hamas ideology will not be able to benefit by causing war.
- The Hamas ideology will be at opposite to the "Palestinian cause".
- There will be incentive amongst the entire Arab world to create an alternative to Hamas.
- Egypt will become responsible for the welfare of Palestinians leaving Gaza.
- The Arab states will become responsible for taking in Palestinians, as Egypt will demand they share the burden.
- The world will no longer be able to blame only Israel.
- The world effort will go from blaming Israel to helping Palestinian refugees.
- The conflict will shift from furthering the Palestinian "cause" to helping individual Palestinians.
- Palestinians will have a choice.
- Hamas will attack Palestinians trying to leave and get bad media coverage.
- IDF will protect Palestinians and get good media coverage.
- Palestinians will get more ideological separateness from Hamas to support new leadership.

Not surprisingly, Egypt is threatening to abandon the Camp David Peace Treaty if Israel opens the Rafah border to let Gazans who wish to leave out. Of course, Egypt has obligations under international law it cannot blame Israel for, and cannot turn away Palestinians (who are refugees, thanks UNRWA) seeking asylum at Rafah.
See: https://blogs.timesofisrael.com/egypt-must-accept-refugees-at-rafah-per-intl-law/

While some may wrongly think this might be ethnic cleansing, the discussion here is only of allowing those Palestinians who voluntarily wish to leave because of the dangerous Hamas-caused war to do so, not forcing anyone. Suspiciously, nations gleefully accept refugees from every other conflict zone worldwide except Gaza, where people are being used as human shields by Hamas.

What might happen long-term as Palestinians who voluntarily wish to leave can go into the Sinai?

- Palestinians will receive food, resources, and care in the Sinai that they cannot receive in Gaza because of Hamas.
- Resources will flood into the Sinai, helping Egypt's economy and development of the largely-undeveloped Sinai.
- The UAE will be pressured to change its $35 Billion land investment bound for Egypt's coast West of Alexandria, and to use the funds to help Palestinians in the Sinai in the East.
- Enclaves of Palestinian areas, towns, and then cities will be built in the Sinai, where there is much more (23,000 square miles of) mostly uninhabited land - many times more Mediterranean coastal land than all of Gaza (141 square miles) Israel (8600 square miles) and West Bank (2100 square miles) that everyone is fighting over.
- The "occupation" will end for all Palestinians voluntarily going into the Sinai.
- The "open air prison" will end.
- The Palestinians will be free to move, travel, visit other places, holy sites.
- Palestinians can travel to other countries and have freedom.
- A Palestinian state can form in the Sinai, as discussed in a 2014 deal Abbas rejected but Egypt's President reportedly approved:
  https://www.timesofisrael.com/egypt-denies-any-plans-to-give-sinai-to-palestinians/
  https://www.timesofisrael.com/lawmakers-back-reported-bid-to-settle-palestinians-in-sinai/

- Palestinians will have their own state, miles of beautiful Mediterranean coastline, a seaport, an airport, a chance to grow their economy bordering on two commercially-successful trading partner states, Egypt and Israel, with complete autonomy and self-governance.
- The Gulf Arab states will be encouraged to help moderate it, so extremists don't spread into their countries.
- Palestinian terrorism will become a regional concern, not all dumped on Israel while others enflame it.
- There will no longer be a zero-sum-game equation where the more land Palestinians get, the less land Israel has. Thereby, the two sides can cooperate toward all mutual goals that don't harm the other.
- Palestinians in Gaza and the West Bank can have a choice to stay there ruled by Hamas or Fatah, or go into the Sinai and have autonomy and self-governance.
- The problem until now has been that "Palestinian cause" is two separate ideological causes actually intertwined, 1) helping Palestinians, and 2) destroying Israel. The problem of the entire conflict is the world cannot help one without helping the other. This solution is to separate the two causes. By opening the Sinai, everything the world does to help Palestinians in the Sinai will weaken the second Palestinian cause of destroying Israel, instead of being a stepping stone.
- Palestinians who want their own peaceful state and an end to occupation win. Israelis who don't want to occupy Palestinians win. The world wins. Hamas loses. Palestinians get virtually everything they want, except closer to destroying Israel.

In other words, the world is opposing the IDF going into Rafah, because it's a potential solution path for much of the conflict. And the world can't have that, now can it.

Maybe we should start chanting, "free free the Palestinians!"

Come on, Israel. Be a light unto the nations.

Seize the damn Rafah border already!

If I time-traveled here from the future, from a time after humanity had learned to solve all our wars, this is what I would tell you about Gaza, today.

Humanity's wars are not over land nor resources nor religion. Wars are not caused by individuals, nor soldiers, nor even leaders for the most part. Wars are caused by the moral questions that humanity stumbles over. When we cannot answer these moral questions, we begin to polarize on opposite sides of them in group collective mindsets. Each side becomes increasingly ideological and irrational, breaking down the systems that maintained peace, believing it is defending itself from a worse threat in the other, until war is the only option left.

That moral question of the entire Israel-Arab conflict in its broadest sense may be phrased as, "What should happen if a displaced, almost-exterminated, unique, minority people reestablish their indigenous homeland after 2000 years, where a non-diverse ideology dominated and is willing to cause its people endless suffering and war to reconquer it?" Unless the world understands a better answer, as humanity naturally desires peace, the peoples of the world will increasingly give in to terrorism, even to demands for Jewish extermination, and even join in, as they have before.

From a future perspective, I would also tell you that the peace-building systems of 2024 are not set up to solve such questions. We have a dispute between the Jewish state on the one hand, and not Islam, but an aggressive ideology within Islam on the other. That ideology wants Israel, the only state that might defend the Jewish people from the next surely coming holocaust, gone. Unfortunately, the world's current peace-building system, the United Nations, has one Jewish state and fifty-seven Muslim states, meaning ideology and politics will skew the world's very sense of justice and morality - the world's very lens through which people understand right and wrong.

Per my article entitled, "The true reason we're not getting the hostages back" (https://blogs.timesofisrael.com/the-true-reason-were-not-getting-hostages-back/) while Israel is killing individual Hamas henchmen, their global anti-Israel ideology is gaining global strength and support. The United Nations' actions against Israel, and recognition of a terrorist-led Palestinian state by over 143 nations show this.

Israel cannot destroy Hamas with bombs, because Hamas is an idea. The only thing that can destroy an idea is a better idea. Otherwise, Hamas will simply reconstitute whenever Israel's military campaign is over regardless whether Sinwar is killed or not, or another group would take Hamas' place, feeding off the same population base ideology. While this ideological problem is worse if Iran and Qatar fuel the conflict, it's not eliminated if they don't.

If the future of Gaza were a question, it might be framed something like: "Why would the Palestinian ideology become moderate?" Because if would not, then they would wage war against Israel no matter what Israel does, no matter how much land Israel might give up. Nations have conquered nations since the beginning of time, not by leaders' intentions, but by ideologies that empowered leaders who did so.

Right now, 72% of Palestinians support October 7th, and in the largest Arab poll taken by the Arab Center for Research and Policy Studies across perhaps sixteen Arab countries, 84% oppose recognizing Israel half a century after major peace treaties with Egypt and Jordan, and after the Abraham Accords. The peace deals between leaders become near-meaningless in light of a raging ideological war inflaming the population base.

For 75 years, Israel has been playing roulette, hoping for a moderate Palestinian leader to emerge, not understanding that the game is rigged, and the ideology will always put in place the more aggressive leader. There is not a single democracy in the Arab world, which itself has no culture of compromise or even a term for "compromise" in traditional

Arabic language, so why would a theoretical "Palestine" magically be the first compromising Arab democracy?

Without this answer, the world calling for a Palestinian state is also calling for the destruction of Israel, knowingly or unknowingly, and thus calling for more war.

And that's the problem in a nutshell. The ideology of the "Palestinian cause" is actually two separate ideologies. One, the "Palestinian cause" of peaceful statehood next to Israel, and two, the "Palestinian cause" of conquering Israel. The root of the problem of the entire conflict is, the two causes are intertwined. The world can't help one without helping the other, making any Palestinian state a steppingstone to destroying Israel, because when you feed an aggressive ideology, it only becomes more hungry, and more aggressive.

The quest for the solution, then, is to answer the question, "How do you help the first peaceful 'Palestinian cause' while also, necessarily, definitely, precisely, weakening the second conquering 'Palestinian cause'?" While Israel cannot solve the entire conflict in one step, any action furthering this dynamic is finally a step in the right direction, and as good as we may hope for.

With no political or democratic process, whatever Hamas does to cause misery toward its genocidal aims, the Palestinians will support, as they have no other choice. Even with Saudi or U.N. governance, it's the same rigged roulette game, and the ideology would just channel all power to the most extreme group, resulting in more of the same. In 1-2 years, they would again lose control of Gaza to Hamas or a Hamas-like group.

But there are solutions.

Without a political process, there's still a way to ultimately hold Palestinian leadership accountable, by holding the ideology itself accountable. And that way is, to allow Palestinians, if they want to leave voluntarily, to be granted access, visas, and asylum, to emigrate to other countries, temporarily or permanently, as they choose.

Why would this help?

Because while dropping bombs kills individual Hamas members but strengthens the ideology, giving Palestinians another option would do the opposite. Ideologies can withstand bombs, but can be weakened if you understand how. If Palestinians can vote with their feet and leave, that weakens Hamas and every extremist ideology that could ever rule, because metaphorically, "extremists can only swim in a sea of moderates". Poor leadership would increasingly cause the metaphorical "tide" to recede by decreasing the supportive population base, the more Palestinians choose to leave Gaza.

When some Israeli minsters discuss voluntary relocation of Gazans, even temporarily, the Arab world fiercely opposes, saying the Palestinian people must not be displaced again, and pressures the world to likewise prohibit all emigration of Palestinians. But this is immoral. This is not the Arab world's decision, but the decision of each individual Palestinian. There is a huge gulf as wide as the Grand Canyon between Palestinians being forcibly relocated, and allowing those who wish to leave voluntarily to do so. Morally, the Arabs cannot win this argument. Since the world agrees on the moral rule of helping refugees escape from every other conflict zone, the world should agree for Palestinians too. The world won't agree, because the world's moral lens is ideologically skewed, now putting the Arab "cause" ahead of individual Palestinians' interests. But if Israel seizes the border and helps Palestinians who wish to go free, Israel will ultimately win the triggered debate that is now being suppressed.

If Donald Trump were to speak to this, he might say, "Not allowing Palestinians freedom to leave is the biggest scam in the world." The Arab world has no right to put its own ideological "cause" which has caused the Palestinian people 75 years of suffering above the Palestinians' own individual right to choose their own futures.

Remember, the goal is not for all Palestinians to leave, but to set up a mechanism whereby the ideology is divided into its two subparts, so Palestinians can best moderate their own governance. By allowing Palestinians who wish to voluntarily leave to do so, and giving them several safe options including in the Sinai, we have set up a reciprocal system where the more conflict their leadership causes, the less fuel for the ideology. The more extreme the leadership becomes, the weaker it becomes. The mechanism lets air out of the entire pressure cooker of a conflict. Hamas cannot claim it is representing the Palestinians if they are leaving. As any Palestinian leadership, Hamas or otherwise, will naturally want more power, it will be finally, after all these years, be incentivized to moderate. Hamas is not afraid of bombs, but is afraid of this. They may make a deal immediately after Israel seizes the border, rather than let one Palestinian leave. Shooting fleeing Palestinians is not a good look for Hamas. Such ideologically-directed strategies are the only thing that can moderate any Palestinian society because they supersede individual, leader, and government interests and decisions, and thus are a prerequisite to any coexistence.

"Not a single Palestinian should be forced to leave, but not a single Palestinian should be forced to stay, either."

This is a moral position everyone should agree on. In a world that has lost its moral compass, unaware of the pivotal moral questions let alone their answers, the one thing the world should be able to agree on, is that the Palestinian individual people should be able to make their own decisions, including putting their own self-interest ahead of the "cause" the world imposes.

Freedom to leave is the real Palestinian "self-determination" the world should be advocating for.

It's not Israel keeping Palestinians in an open-air prison if it is such, but the rest of the world by refusing to accept them.

The Arabs will always put their broader "cause" ahead of individual Palestinian choices, because it's not them suffering the consequences.

Still, even though anyone relocating would be *voluntary*, some will still falsely label it "ethnic cleansing", because buzz words elicit emotions among ideologues easier than intellectual debate resolves them. While "ethnic cleansing" is not a defined crime under international law, a U.N. Commission described ethnic cleansing as S/1994/674 as "… *a purposeful policy designed by one ethnic or religious group to remove by violent and terror-inspiring means the civilian population of another ethnic or religious group from certain geographic areas.*" Israel is not ethnic cleansing, and would not be even if many Palestinians were displaced. Israel would not intentionally use force or terror against the population, Hamas would, and is. Many Palestinians want to leave Gaza voluntarily and have wanted to for years. Additionally, Israel is not selecting people for displacement based on ethnicity or religion, but allowing all people in one conflict zone territory to leave. It's geographic, not ethnic. Palestinians are 99% Arab and Muslim, and are homogenous because they themselves do not allow other religions and cultures, but that's not Israel's fault. So, if a group that is homogenous because of their own discrimination is displaced, their discrimination does not transfer onto Israel. In fact, Israel has two million Arabs, so if under Israeli control, Gaza would have Israeli Arabs too, and be *more* diverse.

Egypt has allowed 50 tunnels from the Sinai to Gaza through which Hamas and weapons can enter, instigating this conflict and causing untold Palestinian suffering for decades. The tunnels found alone give Israel the moral right to seize the border.

But Egypt sealed its border and is also in violation of its obligations under international law to give asylum to any Palestinian refugees at the Rafah border. https://blogs.timesofisrael.com/egypt-must-accept-refugees-at-rafah-per-intl-law/ This is a whole additional justification for Israel to seize the Rafah border to help Palestinians who wish to flee the conflict zone to do so.

If the world blames Israel for food and aid Hamas steals, Israel has yet another ground to seize the border, establish a safe zone, and deliver food and aid to Palestinians in the Sinai where Hamas is not in control.

Israel could bring a case before the International Court of Justice against Egypt for helping instigate this war, and violating its obligations to accept refugees, demanding not only confirmation of the law, but $50 billion in compensation for all the suffering Egypt caused the Palestinians, and might resolve the war before the case came up for trial.

Israel is afraid of Egypt's threats to withdraw from the 1979 Camp David Peace Treaty, without understanding that Egypt has helped put Israel into an existential war, and is blocking key peace solutions. Under the treaty, Israel has the right to "agree on the modalities for establishing the elected self-governing authority in the West Bank and Gaza." Egypt violated this by allowing tunnels and weapons in.

Israel has a right to "arrangements for assuring internal and external security and public order", which Egypt violated. Israel also has a right to "participate in joint patrols and in the manning of control posts to assure the security of the borders." Israel also has a right to have Egypt take "All necessary measures will be taken and provisions made to assure the security of Israel and its neighbors during the transitional period and beyond" and Egypt violated that as well. The Treaty also says, "Egypt agrees that the use of airfields left by Israel near El Arish, Rafah, Ras El Nagb and Sharm el Sheikh shall be for civilian purposes only, including possible commercial use by all nations." So, Israel has a right to use the airfield to bring aid to the Palestinians. Israel has exhausted its remedies in seeking a negotiated solution with Egyptian President El-Sisi under U.N. Charter Article 33, but Egypt refuses to negotiate. Israel is being baited with a Saudi peace deal, right into a continuing or worsening conflict as no deal would change the Arab world's underlying ideological dynamics. So, Israel must help Israelis and Palestinians, and be a light unto the nations by taking a moral position. And, Israel should invite International media and monitors to hear, see, and debate as Palestinians have the option to flee into the Sinai for a better life than Hamas offers, so the rest finally have a chance to stay and make peace.

Five peace offers didn't work.

Elections didn't work.

Cutting off aid to the Palestinians didn't work.

Giving aid to the Palestinians stolen by Hamas didn't work.

Maybe "free Palestine" should be "free the Palestinians!"

Seize the damn Rafah border already!

THE END

Thank you for reading.

For more, please subscribe to my Substack @

www.danielbenabraham.substack.com

Please read my Times of Israel Blog at:

https://blogs.timesofisrael.com/author/daniel-ben-abraham/

Please see my sites at:

www.danielbenabraham.com

www.thepeacematrix.org

Follow me on Twitter @thepeacematrix & @danielbabraham1

Appendix A: List of 26 Starting Question categories

Here are the PeaceMatrix™'s 26 categories for solving every human conflict possible, ever existing and forever to exist:

A INTRODUCTION – What is the best summary and scope of the dispute for peace?

B PARTIES AND TERMS – What are the best understandings of important definitions and party identities for peace?

C HISTORY AND CURRENT SITUATION – What is the best understanding of the history and current situation for peace?

D WANTS – What is the best understanding of the full spectrums of each party's wants for peace?

E DISPUTE – What is the best understanding for peace of why the parties' wants are problematic and where the dispute actually lies?

F COMMUNICATION – What is the best understanding of communication involving the parties for peace?

G UNKNOWN – What are the most important things each side does not understand that would be helpful to know for peace?

H CULTURES / IDEOLOGIES – What about the parties' values, cultures, human natures, animal natures, beliefs, ideologies, customs, national, local, and tribal interests and other unknown forces and motivations is most important for understanding and peace?

I WRITINGS – What are the most important understandings about the relevant key writings, documents, rules and laws for peace?

J MORALS – What are the most important understandings about the parties' different subjective moral codes for peace?

K MORALITY BUILDING – What would and should the parties' ideal culture, morals, values and rules be for peace, shared or individual?

L WHY NOT IMPROVE – What is most important to understand for peace about why each side might want or not want improvement or a resolution?

M LEADERSHIP – What is most important to understand for peace about the key authorities and decision makers?

N SCIENCE, TECHNOLOGY, AND RESOURCES – What are the best understandings of all science, technology, and resources necessary for peace?

O COMMON INTERESTS – What is the best understanding for peace of the common ground, common interests, common goals, common threats and common enemies other than each other?

P OBSTACLES – What is the best understanding for peace of obstacles facing the parties?

Q CULTURE-BUILDING – What common culture can the parties build together which preserves their own and also unites them?

R COMMUNICATION BUILDING – How could the parties better communicate for peace?

S WHY RESOLVE – What might happen is the dispute is not resolved, and what are the motivations for seeking peace?

T PAST SUCCESSES – How have similar challenges been attempted to be resolved, improved, or actually resolved, elsewhere and in the past?

U SELF-IMPROVE – What can the parties do to unilaterally improve their positions without harming the others'?

V WORKING BACKWARDS – What are ideal future scenarios for peace, and how can we work backwards from these? If the dispute would have been resolved by a future point but for certain factors, what are they?

W MORE PARTIES – Who else can be brought to the negotiating table and how would they help peace?

X SOLUTIONS – What are all possible peace proposals, ideas, and solutions?

Y NEW VALUES – What rule, doctrine, or morality is created by each positive scenario and possible peace outcome?

Z CHANGES – What changes/additions should be made to this PeaceMatrix™?

Read Volume I if you haven't already for more about how the system is intended to work.

www.ingramcontent.com/pod-product-compliance
Lightning Source LLC
Chambersburg PA
CBHW062219270326
41930CB00009B/1789